AquaGuide

Freshwater Stingrays

Hans Gonella
& Dr Herbert Axelrod

CONTENTS

First published in the UK in 2003 by Interpet Publishing,
Vincent Lane, Dorking, Surrey RH4 3YX, England

English text © 2003 by
Interpet Publishing Ltd.

ISBN 1-84286-083-6

The recommendations in this book are given without any guarantees on the part of the author and publisher. If in doubt, seek the advice of a vet or aquatic specialist.

Translation: Heike de Ste. Croix
Technical consultant: Richard Hardwick

Originally published in Germany in 1999 by bede-Verlag,
Bühlfelderweg 12,
D-94239 Ruhmannsfelden
© 1999 by bede-Verlag

Picture Credits
The pictures were supplied by Dr Herbert Axelrod, Bernd Degen/bede-Verlag, Hans Gonella, Dr Jürgen Schmidt/bede-Verlag, Yvette Tavernier/Animaqua, as is credited in the captions.

Freshwater stingrays are fascinating fish to keep in a home aquarium. At the same time they demand a considerable degree of responsibility from their keeper who must provide the right environment and care for this naturally graceful fish. Although freshwater stingrays are generally easy to keep, they have certain requirements which, if they are not met, can be detrimental to their health. Unfortunately, even today many freshwater stingrays die because they are either kept in cramped conditions or because of poor water quality.

Freshwater stingrays are often considered to be the superstars of the fish world. Because they are relatively expensive to buy, it is easy for an aquarist to assume that he has bought something very special, a purchase which will raise the status of its owner. Or, to put it another way, because of its venomous sting, people are eager to experience the "ultimate thrill" in home aquatics. However, neither of these emotions do these beautiful fish any justice. They require the maximum amount of attention from their keepers and their commitment to provide an environment and regime of care appropriate to their needs. The normal body size of adult rays alone, which is between 60-80cm, gives an indication of the size of aquarium required. Furthermore, freshwater stingrays are very lively creatures which, if kept in cramped conditions, soon lose their attraction. Nothing is more tragic than to watch a healthy, strong, vivacious ray trying to move freely in a small aquarium. Sooner or later, the frustrated aquarist will have to find it a new home.

This book has been written to cover all aspects of the correct care of rays in order to make it possible for fishkeepers to keep them successfully. By reading this book attentively, the aquarist who is eager to learn will soon discover that our knowledge of freshwater stingrays is not nearly as comprehensive as one might assume. Much is still unknown and many

Riverbanks – the Rio Negro is pictured here – are the preferred hiding places of stingrays.
Photo: bede-Verlag

mysteries have not yet been uncovered by science.

In addition parts of the systematics are incomplete, a fact which becomes apparent when rays are continually imported that cannot yet be classified according to their species. Indeed the descriptions of some species are pure

The underside of the ray is light in colour as it does not have to be camouflaged.

speculation as definitive knowledge with regard to determining the species is not available.

Certain facts are repeated several times in this book, specifically in those places where they are relevant, so as to help the reader to understand interdependent aspects of stingray care. So it should be possible to use these explanations as a reference in order to find answers quickly to specific questions.

Notwithstanding these warnings, and the fact that rays make considerable demands on their keepers which must not be underestimated, anyone interested in these fascinating freshwater fish should not be dissuaded from studying them in detail. They will be rewarded with a fantastic and gratifying experience.

This picture of the underside of a female ray shows the gill slits clearly. Two spiracles, positioned directly behind each eye, enable the rays to breathe even when buried in the substrate.
Photo: Yvette Tavernier

"Rays are flat fish which inhabit the oceans." This is the definition often given by fleeting observers. As a matter of fact, most ray species do live in the sea. The most commonly known and most frequently seen on many TV documentaries are the manta rays because of their remarkable wing span, followed by the families of the electric rays, eagle rays, guitarfishes and skate. Sawfishes are also relatively well known; and let's not forget the stingrays, which live in fresh water as well as salt water.

Most aquatic fish are bony fish; in total more than 20,000 bony fish species are known. Rays on the other hand are cartilaginous fish, which are a relatively small group comprising approximately 1000 species. Among vertebrates, rays form their own group of cartilaginous fish. Sharks, rays and catfish are all cartilaginous fish. Sharks and rays are related. Some ray species are similar in appearance to sharks and vice versa.

As the name cartilaginous fish indicates, the skeleton of this group of fish consists almost entirely of cartilage. The skeleton has no bones whatsoever. It is very flexible, yet tough enough to guarantee the necessary stiffness. Calcium deposits are only found in the cartilaginous vertebrae and in some other parts of the skeleton. The scales and teeth also contain some bony particles.

Sharks and rays belong to the elasmobranch sub-class of the Chondrichthyes class of cartilaginous fishes. They live in fresh water as well as salt water. Ratfish (or chimaeras), however, have never been found in fresh water. The elasmobranch sub-class is split into two groups. Alongside the few "real" freshwater cartilaginous fish live many other cartilaginous fish that were originally marine varieties but which are sometimes found in fresh water. The great majority, however, are only found in salt water. Only freshwater stingrays and freshwater sharks have completely adapted to life in fresh water. All other sharks, sawfish and rays, which are occasionally found in rivers, return at least for the breeding season to the ocean.

The natural habitat for rays extends across the whole globe. Rays inhabit all oceans and live up to 3000 metres below sea level. Rays live in temperate climatic zones as well as in the sub-tropics and tropics, and they are mostly observed by divers in shallow coastal waters. Some ray species spend part of their lives in brackish water, i.e. in those regions where big rivers flow into the sea bringing large amounts of fresh water with them. This reduces the salt content considerably in the respective coastal regions. Some species move from the brackish water into fresh water and occasionally swim large distances upriver, where they spend part of their lives. The change from salt to fresh water requires a considerable physical transformation. How the rays, and indeed the sharks, achieve this is one of the many mysteries surrounding elasmobranch fish.

Freshwater stingrays

Stingrays form their own family of fish, which among other types consists of the genera *Dasytis* and *Himantura* as well as the *Potamotrygon* species. The latter are the real South American freshwater stingrays. They are also the only freshwater stingrays which can be kept permanently in a home aquarium. The Asian freshwater rays of the genus

Himantura grow to such extreme body circumference that they are completely unsuitable for home aquatic purposes.

This book is therefore limited to considering the more frequently encountered South American freshwater stingray species, the potamotrygonids, although it does contain some additional general information concerning the biology and evolution of rays for better understanding. Information on the proper care of rays in a home tank is only available from experience gained with South American freshwater stingrays. It has not yet been established how much of this can be applied to other freshwater stingray species. In any case, most freshwater stingrays imported are from South America, with the majority coming from three species: *Potamotrygon motoro*, *P. leopoldi* and *P. orbignyi* (the latter species is also known by the name *P. reticulatus*). Therefore all recommendations regarding care are confined to these three species. However, from the latest aquatic knowledge, it seems fair to assume that the care requirements for other South American freshwater stingrays are not much different.

Shape, body functions and anatomy

The shape, physiology and anatomy of the South American freshwater stingrays do not differ much from those of other ray species. The appearance is typical for rays, namely like a flat disc.

They glide through the water with wavy movements created by the fin margins. Propelled only by both pectoral fin margins, they can move at an astonishing speed, while the body axis remains static.

The reason for this is the stiff spine, which is found in most ray species, and which influences the way they move.

The pectoral fins are constructed in such a way that they almost border the whole body. Right at the base of the tail underneath the body is the pelvic fin. The anal and dorsal fins are missing completely. Instead the South American freshwater stingray has a tail fin at the end of its long tail. Immediately next to the pectoral fins, left and right next to the cloaca, the males have two secondary sexual organs, or claspers, which are evolved extensions of the pectoral fins. The pipe-like extensions are for internal insemination during which the male inserts one of the organs into the cloaca of the female. This keeps the loss of sperm to a minimum. The secondary sexual organs of the male are also used to determine the sex. They are already clearly formed in newly born males. The reproductive organs of some ray species are still small and in others they are quite long. At maturity the reproductive organs reach a remarkable size.

All known South American freshwater stingrays have a row of more or less distinctive sharp barbs on their tails. The continuation of this is found, as it were, in the ray's sting, which is covered with a fine layer of skin. Quite often a second sting is already developing under the first one, which in time will replace it. Sometimes even a third sting is already hidden under the first two. In mature rays the sting is really only replaced once a year, whereas young rays replace their sting several times a year. The tail sting as well as the sharp barbs are nothing but enlarged or modified placoid scales.

The ray's sting is flat and serrated on both sides. It is firmly lodged in the tissue and is a very dangerous defence weapon. It is so hard that, with one strong tail movement, it can even penetrate wood. It is not only dangerous because of the wound it inflicts by leaving a sting behind, but primarily because of the venom which is released on impact. The venom is produced at the base of the sting. The venomous cells which produce the poison are apparently supplemented by two additional poison glands which are connected to nearby mucus cells. These are responsible for transporting the venom to the sting. The venom collects under the fine layer of skin around the sting and remains there until it is released. It is quite possible that the distinctive jagged edges help to release the venom. As soon as the sting is used as a defensive weapon, the fine layer of skin breaks on the jagged edges so that the venom can be released. The composition of the venom has not yet been sufficiently researched. Among other substances it contains a so-called serotonin in conjunction with free amino acids, enzymes and other proteins which cause severe poisoning.

The scales of the ray are a special feature, which feel like coarse sandpaper. Firmly embedded in the skin these scales are called placoid scales. The base plate of one scale is anchored into the skin on top of which lies the denticles. The base plate consists mainly of bony material. The spines of placoid scales consist of dentine. Their surfaces are covered with enamel. The placoid scales are formed differently depending on the species. They are spread unevenly over the body, with some parts of the body, especially the belly, completely free of scales. The

numerous small placoid scales protect the rays' bodies from accidental damage.

The teeth of rays are linked by evolution to the placoid scales. Furthermore, the toothed jaws of vertebrates are considered a development of placoid scales. Hardly visible externally, the ray's teeth lie immediately behind its mouth opening. South American freshwater stingrays have flat grinding teeth. They produce so-called renewable tooth plates, whereby several rows of teeth from the upper and lower plates are used simultaneously. If teeth are lost, the gaps are quickly filled by newly grown teeth which literally line up at the front. This constant tooth replacement enables the rays to feed on crustaceans and the wear and tear on their teeth poses them no problems.

The eyes of rays are slightly raised on the upper side of the body. This gives them

Rays have exceptional vision. Even at night with only very little light, they are aware of their surroundings. Photo: Yvette Tavernier

Freshwater stingrays need large tanks. Photo: bede-Verlag

a wide field of vision. The eyes are very agile and can, if required, be retracted. The eyelids move across the eyeballs which, when the rays submerge into the substrate, prevents dirt getting into their eyes, and thus their vision is not impaired. Rays have good vision even in the dark. Small mirror plates consisting of guanine crystals, which are situated in the eyes, reflect the smallest amounts of light in the dark and thus stimulate the visual cells many times more than normal. This particular feature, which is also found in cats, can be observed in fast fading light.

The spiracles are easy to spot behind the eyes. These spiracles are modified gill slits. These allow the rays to inhale water even when submerged in the substrate. As soon as it submerges, the ray clears its spiracles and is able to inhale water through these openings which is then transported into the gills.

After that the water is then exhaled through the gill slits on the ventral surface. These two rows of five gill slits can be compared with the gill covers of bony fish.

Tip: Rays have good vision and a well-developed sense of smell.

The nose openings are connected with the mouth opening, each divided by a fold of skin to allow water to flow in and out. However, rays can really only smell well while swimming. In addition, they have a sense of taste as well as a sense of touch which helps them to find food. It is fair to assume that the South American freshwater stingray – like all elasmobranch fish – also has an additional power of perception (see box below).

Tip: Their "electrical sense" helps them to detect electric impulses emitted by muscle contractions in their prey. This enables them to find food even in complete darkness.

The South American freshwater stingray has no swimbladder to regulate its body movements. This means that rays sink to the bottom as soon as they stop swimming. Instead of the swimbladder, the relatively long intestine and the conspicuously large liver together with the other organs take up what limited room there is inside the abdomen. From the design of the intestine we can conclude that rays mainly eat a vegetarian diet which is low in nutrients.

The water balance of rays is regulated in a particular way. Members of the elasmobranch group which live in the sea store large amounts of urea in their blood and tissues. This gives them an isotonic balance to the sea water and they hardly lose any water through their skin. Marine elasmobranch members therefore do not drink but absorb water through their gills. Freshwater stingrays on the other hand have hardly any urea in their blood. This quantity would not even be sufficient to regulate the osmotic water balance of diluted salt water. The reason for this is the anal gland, which in South American freshwater stingrays has no function, which means it no longer works. Therefore it is no longer possible for these freshwater stingrays to live in the sea, where they would soon die.

Evolution of freshwater stingrays

Rays belong to one of the oldest animal groups. They were already an evolutionary successful group of fish when dinosaurs ruled the world. Presumably cartilaginous and bony fish from the early Palaeozoic had common ancestors. Very little is known, however, about the appearance of the very first cartilaginous fish. The cartilage skeleton is not very durable and disintegrates quickly. Only the scales and teeth of rays survived to be fossilized and they give only a very modest insight into their evolution. The first findings of cartilaginous fish came from Silurian geological layers dating from approximately 420 million years ago. Later in the Permian, approximately 280 to 225 million years ago, ancient sharks evolved into the modern forms.

Some 220 million years ago the very first rays appeared. However, fossil finds of ancestors of today's living rays only date back to the Jurassic period approximately 190 million years ago.

At about the same time when rays were spreading through all the oceans of the world in the Palaeozoic and Mesozoic, mountains began to form in South America. In those millions of years, what is nowadays known as the Andes were created. As the land slowly rose and water levels sank, many sea-living creatures moving into fresh water were faced with slowly changing living conditions. As mountain ranges formed, waters began flowing in different directions. Those waters which had previously flowed into the Pacific were blocked and had to find new river beds. What are now the waters of the Amazon therefore had to change direction and they now flow into the Atlantic. During this continual process fish

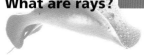

Rays are close relatives of sharks. Some sharks are even similar in appearance to rays and vice versa.
Photo: Hans Gonella

which lived at least some of the time in brackish and fresh water found it more and more difficult to return to the Pacific, until it was no longer possible. Some of these species, however, adapted to these new living conditions and former marine creatures evolved into freshwater ones. One of them was the ancestor of today's freshwater stingray.

From what we know today these freshwater stingrays are descendants of marine rays with a round disc-like form. Over millions of years the shape and in some respect also the anatomy of rays changed sufficiently to enable them to evolve into freshwater fish. The descendants of these formerly marine rays can still be found today in the East Pacific and the West Atlantic. It is quite probable that the ancestors of the South American freshwater stingrays were species of *Urolophus* and *Urotrygon*, which evolved in a different way.

Habitat
Today South American freshwater stingrays inhabit almost all waters of South America. The potamotrygonids, more than all other freshwater elasmobranch members, have found the largest habitat. The waters of the Orinoco, the Amazon and the Rio Paraná are their main habitat, but South American freshwater stingrays are also found in Guyana and Maranhão as well as regions of the Mato Grosso.

Rays are mostly seen in shallow waters near sandbanks; however, the great majority of them live in almost all tributaries of the aforementioned stretches of water. They are comfortable in deep water as well as near shallow river banks. They even roam the flooded areas of the rain forest. The largest population of South American freshwater stingrays is probably found in the calm water of lagoons and river backwaters. These habitats have thin layers of mud which provide an abundance of food. Apart from salt water in which rays cannot live, rapids are the most significant natural barriers which separate individual ray populations. However, whether this has any influence on where individual species are found has not yet been established. Based on only a few underwater observations, we can assume that rays roam large areas and do not only live in shallow zones by the banks.

The legend of rays
In their native countries South American freshwater stingrays are feared more than the infamous piranhas. Locals insists on warning foreign tourists especially about

coming into contact with rays. The reason being not so much that rays attack humans, but a leisurely walk through shallow waters could lead to a chance encounter with a ray hidden in the sand. Because of its camouflage, the ray has no reason to flee. Piranhas on the other hand make a quick escape as they are generally cautious about coming into contact with larger objects. Rays will remain motionless until the human touches them. Stepping onto or even next to a ray is a disaster. As quick as lightning, the stingray will try to fend off the

stay in the water. They shuffle through the shallow water by slowly putting one foot in front of the other and in doing so pushing the sand or mud on the river bed ahead of them. This way they avoid treading on a ray and instead any ray

South American freshwater stingrays are feared more in their native countries than piranhas or caymans. Photo: Daniel Lüthy

***Below:** This is the fossil of a guitarfish, Rhinobatos beurleni from the Lower Cretaceous of Brazil. Fossil remains of cartilaginous fish are rare as their skeletons decay relatively quickly. Photo: Hans Gonella*

hidden in the substrate is carefully pushed to one side if it hasn't made its escape already. This greatly reduces the risk of coming into the contact with stingrays.

> **Tip: A sting from a ray can have serious consequences. Quite often the sting, which is covered in small barbs, breaks on penetration and remains in the wound.**

supposed attacker with a powerful tail movement, whereby it can lash its tail forward just like a scorpion and inflict dangerous injuries with its sting on the hapless person. The ray then disappears immediately and is not seen again. In order to avoid such accidents, locals who work in and around the waters have developed a simple method which enables them to

It is especially dangerous when the sting hits a bone. The blow of a ray's tail is so powerful that the sting can easily penetrate rubber boots.

Unfortunately in some parts of South

America freshwater stingrays are hunted without mercy because of their unpopularity. They are speared and carefully hauled onto the boat where their tail bearing the sting is immediately chopped off. Quite often, the animal, while it is still alive, is thrown back into the water. Occasionally, a ray will survive this horrendous action which is why some rays without tails have been spotted in rivers. The stings are not discarded but sold as they are much prized in the Amazon regions. The stings are sometimes sold to tourists but more often they are used by locals for magic rituals.

Europeans are usually not aware of the danger of the ray's sting. The observer is merely fascinated by the appearance of the stingray. This reaction is different in South America. There are unsubstantiated reports that even the Amazon Indians hack off the tails to avoid inadvertently coming into the contact with the venomous sting. The meat of rays is nowadays no longer important for the locals. Ray meat is hardly ever sold at markets. This was not always so. When the first discoverers and missionaries sailed in South American waters ray meat was a staple diet for many Indian tribes. Those rivers in which populations of stingrays were found were literally called stingray rivers by the Indians. Apart from the meat, the Indians also valued the hard sting, which they used as an arrowhead. It was also used as a tattoo needle or for piercing arms, ears and tongues so that they could wear ritual objects or jewellery.

It is assumed today that the Indians invented the shuffling walk to allow them to go fishing safely in the shallow waters. As some of the Indians were occasionally stung, they developed all sorts of alternative medicines to treat the wounds. A variety of plant substances were used as treatment. Unfortunately, only very few traditional Indian tribes still live in remote regions and therefore an enormous amount of medicinal knowledge has regrettably been lost forever.

South American freshwater stingrays

New unidentified South American freshwater stingrays are increasingly being sold in aquatic shops. In most cases it is almost impossible to identify them by their appearance. On the one hand some species are very variable, which means that colour and colour intensity as well as markings can be completely different in one species. On the other hand colour or markings can change with age. Even their state of health can have an effect on their colour. With some species, e.g. *P. orbignyi*, the conditions in which they are kept can lead to a change in their colour. If kept in a brightly lit tank with light sand, their body colour will change to a light beige. If kept in darker conditions, their colour changes to dark brown.

The reason for the changeable and inconspicuous body colour of South American freshwater stingrays is easy to understand. Their appearance is for most species their camouflage. The better they can adjust to their habitat, the more protected they are. Conspicuous colours or markings would only spoil this camouflage. Lack of details about their exact origin also makes the identification of species quite difficult. From time to time some species look very similar so that at best one can only guess to which species they belong. Up to now science has only examined South American freshwater stingrays quite superficially. The species described briefly over the following pages therefore only represent a small proportion of the actual vast numbers of extant species. It is quite possible that some ray species are still waiting to be discovered. Occasionally, such species appear in shops which only adds to the confusion surrounding species identification.

To what extent the individually known species are suitable for care in a home aquarium has not yet been established. It is quite possible that the expected size of the ray determines success or failure of permanent care. Experience shows that the most commonly available species, such as *Potamotrygon motoro*, *P. orbignyi* and *P. leopoldi*, do have a limited suitability for home aquatics provided the tank is of adequate size.

One must expect that some species may grow to 1m and larger, and yet the fully grown Ocellated Freshwater or Peacock-eye Stingray, *P. motoro*, which has a disc diameter of 80cm, is considered a real giant among aquarium fish. The

Identifying the species of many rays is very difficult, if not impossible. Among other factors, the variability of certain species contributes to this. This is most probably an unidentified species. Photo: Hans Gonella

black rays of the *P. leopoldi* species, however, do not grow bigger than 40cm in diameter.

Systematology

All natural species are classified in a hierarchy system. The group of cartilaginous fish belong to the super-class of vertebrates with jaws (Gnathostomata). In the class of cartilaginous fish, sharks, rays and skates are all grouped together. Within the order of the rays, several sub-orders exist, such as the sawfishes, electric rays, guitarfishes, skate as well as the stingrays. The freshwater stingrays belong to the latter grouping. The Dasyatidae and Potamotrygonidae families comprise the ray species which live exclusively in fresh water. This was not always the case however – but more about that later. The Potamotrygonidae family contains the three ray genera living in South America. The genus *Paratrygon* and the genus *Plesiotrygon* each only have one species. The third genus *Potamotrygon* contains around 18 species which are known today. But it is quite possible that in the future other species will be discovered which also belong to this genus.

Unfortunately, freshwater stingrays are not regularly classified by experts so there exist quite a few discrepancies regarding the various species. At the same time "unknown" species are continually introduced without being classified into a species group. This why people have started to number unknown South American freshwater stingrays in Japan. This trade number code has also been adopted in Europe. As a result the ray species of the *Potamotrygon* genus are numbered P1, P2 and so on without there

being any certainty that some of the fish with different numbers are not actually one and the same species.

In Japan the classification of the sub-order Myliobatidoidei was revised after a study by NISHIDA in 1990. This also changed the valid classification of the South American freshwater stingrays that had been used up to that date. With this revision the Potamotrygonidae family was cancelled and transferred to the Dasyatididae family although with reservations about the genera *Paratrygon* and *Plesiotrygon*. As this review was based on a scientifically valid paper the genera *Potamotrygon, Taeniura, Dasyatis* and *Himantura* now belong to the Dasyatididae family. The family classification mentioned at the beginning of this chapter is therefore no longer valid. It may, however, be possible that this study will be revised again because, as we have mentioned before, there exist certain doubts and reservations.

Although the scientific classification of freshwater stingrays is of little importance to the aquarist, it has been included here to complete the picture and also to prevent possible confusion arising in the minds of the reader.

Ray species

The following species classification is based on the dissertation and species list prepared by Dr Ricardo de SOUZA ROSA in 1985. This work has unfortunately not yet been published and is therefore not valid taxonomically! This species classification is nonetheless very valuable as no other comparable works exist. It is useful for aquarists as it helps them identify South American freshwater stingrays in other

From what is known to date, a precise classification of species is not of vital importance for care of stingrays in a home tank. It is, however, a shame for aquarists that species classification is often not possible.
Photo: Hans Gonella

publications. Consequently aquarists can compare and evaluate scientific data on ray species with other publications.

Individual ray species vary greatly in respect of their colour and markings. The following descriptions of rays are therefore a humble attempt to characterize the species. Colours and markings can also change dramatically with age. This variety in appearance is part of the ray's camouflage; colour and markings even change depending on their habitat, so that fish of the same species can vary from a light, sandy colour without any visible markings to a dark brown colour with pronounced markings. To what extent the different ray species are hybridized has also not been clarified. In 1965 based on

their observations of over 2000 rays CASTEX & MARTINEZ ACHENBACH posed the question as to whether *P. labradori* might be a cross between *P. motoro* and *P. falkneri*.

Many species cannot be identified by their appearance alone. Especially when reliable information about their origin is not available, species classification can only be assumed. Furthermore, large numbers of unknown species have been captured, some of which can surely be assigned to one of the known groups. Others, however, cannot be identified. Following the species compilation we have therefore listed the "nameless" species of rays with the numbers P1 to P61. This list comes from the "Freshwater Stingray Ident-ification Guide" by Richard Ross.

Genus *Plesiotrygon:*

Plesiotrygon iwamae ROSA, CASTELLO & THORSON, 1985
Synonym: (This species is often mistaken for *Paratrygon aiereba.*)
Habitat: The Amazon in the regions of Manaus in Brazil and Rio Napo, Equador.
Colour: The base colour varies between a pale yellow-brown to a grey-brown. The markings are made up of a number of small, narrow black spots as well as many small white ones, which form rosette-like patterns. The markings decrease towards the edges. The colour of young rays is lighter and the spots appear larger.
Special Features: This species lives in deep rivers. Even disregarding the remarkable whip-like tail, their body is considerable larger than 50cm. These rays are therefore unsuitable for home aquatics.

Genus *Potamotrygon:*

Potamotrygon brachyura (GÜNTHER, 1880)
Synonyms: *Trygon brachyurus, Potamotrygon brachyurus, P. brachyura, P. brumi, P. bruni, Ellipesurus brachyurus, Paratrygon brachyurus.*
Habitat: North-east Argentina, Western Brazil (Mato Grosso), central Paraguay and western Uruguay.
Colour: The base colour is brown. The markings in the body centre and on the tail consist of wide, as well as circular, net-like patterns. These are missing from the edges of the fins.
Special Features: Nothing validated is known about caring for them in an aquarium.

Potamotrygon castexi CASTELLO & YAGOLKOWSKI, 1969
Synonym: n/a.
Habitat: North-east Argentina, Central Paraguay, eastern Bolivia and eastern Peru, as well as western Brazil.
Colour: It varies from light-brown to dark-brown. Many small, circular and kidney-shaped yellow spots form the speckled markings. In young rays these markings are arranged in winding lines. In adults the spots form a rosette-like pattern. Some specimens have yellow and black eye-spots on the fin margins. Occasionally, individual black spots can be seen.
Special Features: Nothing validated is known about caring for them in an aquarium.

Potamotrygon constellata (VAILLANT, 1880)
Synonyms: *Taeniura constellata, P. circularis, Paratrygon circularis.*
Habitat: The Amazon, Brazil.
Colour: Brown to pale grey-brown base colour with irregularly spread white spots on the fin edges. Some specimens have dark, net-like patterns, which are most prevalent in fish with a light yellow colour.
Special Features: Nothing validated is known about caring for them in an aquarium.

Potamotrygon dumerilii (CASTELNAU, 1855)
Synonyms: *Trygon dumerilii, Taenura dumerilii, Paratrygon dumerilii, Ellipesurus dumerilii, P. dumerilii* (GERMAN, 1877).
Habitat: Rio Paraná and Rio Paraguay as well as the Rio Araguaia in Brazil.
Colour: This species has a light-brown

base colour with dark, more or less uninterrupted lines which form a net-like pattern. In addition some specimens also have dark spots. Within the net-like pattern, triangular yellow-brown spots complete the body markings. Spines are arranged in an irregular line on the tail.

Special Features: These rays are often mistaken for similar-looking species.

Potamotrygon falkneri CASTEX & MACIEL, 1963

Synonyms: *P. vistrix. P. hystrix, P. menchacai.*

Habitat: North-east Argentina, central Paraguay and west Brazil.

Colour: The typical body colours consists of a dark-brown base colour covered by many irregular oval or kidney-shaped yellow or white spots. These rays have a relatively short tail. The skin of the young rays is smooth and the tail spines are missing.

Special Features: These rays also live in fast-moving waters.

Potamotrygon henlei (CASTELNAU, 1855)

Synonyms: *Trygon henlei, Taenura henlei, Taeniura henlei.*

Habitat: Northern Brazil in the Rio Tocantins and Rio Araguaia.

Colour: These rays have yellow or orange "eye-spots" on a dark, olive-brown to dark-brown to almost black background. Every now and then black flecks are found on the light spots. The light spots can also have blurred edges. The spots on mature fish are very irregular.

Special Features: It is presumed that these rays only live in the aforementioned rivers. The species looks very similar to *P. leopoldi.*

Potamotrygon histrix (MÜLLER & HENLE, 1934)

Synonyms: *P. hystrix, Trygon histrix, Trigon histrix, Taeniura hystrix, Paratrygon hystrix, Ellipesurus hystrix, E. histrix.*

Habitat: Rio Paraná, Argentina.

Colour: The dark-brown body colour can sometimes show hints of violet-coloured specks. A variety of winding dark lines create a different appearance in some specimens. The fin edges are dotted with many small light spots. The tail is covered with two or three rows of spines.

Special Features: These rays are said to live a solitary life outside the breeding season.

Potamotrygon humerosa GARMAN, 1913

Synonym: n/a.

Habitat: Northern Brazil, Rio Tapajós to Rio Pará.

Colour: This species has a dark net-like pattern on a light-brown to brown background.

Special Features: The relatively long sting is conspicuous.

Potamotrygon motoro.
Photo: Yvette Tavernier

Potamotrygon histrix.
Photo: Yvette Tavernier

Potamotrygon leopoldi CASTEX & CASTELLO, 1970
Synonym: n/a.
Habitat: Brazil, Rio Xingu as well as Mato Grosso.
Colour: The base colour is black. White or yellow spots irregularly spread over the body give this species its typical pattern. Occasionally the spots can have a black centre. Sometimes white interrupted circles appear. Three rows of spines are arranged on the tail. The head of this species is usually a little narrower than that of *P. benlei*.
Special Features: *P. leopoldi* and *P. benlei* are frequently mistaken for one another.

Potamotrygon magdalenae (VALENCIENNES, 1865)
Synonym: *P. magdalenae* (not VALENCIENNES), *Taeniura magdalenae*, *Taenura magdalenae*, *Paratrygon magdalenae*.

Habitat: Rio Magdalena and Rio Atrato, north Columbia.
Colour: The base colour is light brown to brown. Mature adults have a very dark brown colour. Yellow spots arranged irregularly appear towards the edges. Occasionally these spots are disposed in snake-like lines. However, they disappear completely at the edges. In mature adults the spots lose their colour intensity.
Special Features: Nothing validated is known about caring for them in an aquarium.

Potamotrygon motoro (NATTERER, 1841)
Synonym: *Raja motoro, P. motoro* (not NATTERER), *Potamotrygon laticeps, Taeniura motoro, Trygon mulleri, Ellipesurus motoro, Potamotrygon circularis, Paratrygon motoro.*
Habitat: The Amazon and Orinoco.
Colour: The base colour is olive-brown

to dark grey-brown. Irregularly arranged yellow or orange eye-spots and many circular spots line the body's edge. In some specimens the eye-spots are missing from the middle of the body and the tail.

Special Features: The Peacock-eye Stingrays are surely the best-known of the South American freshwater stingrays. Their bodies can reach a diameter of 80cm.

Potamotrygon ocellata (ENGELHARDT, 1912)
Synonym: *Trygon hystrix ocellata.*

Habitat: The Amazon of Mexiana and Mecapá.
Colour: The base colour is olive-brown to brown. The fish have dark-orange to rusty-red irregularly spread eye-spots. In gen-

eral the eye-spots have a dark edge and decrease towards the edges. In contrast to *P. motoro*, the eye-spots of *P. ocellata* have an irregular shape. A row of spines is arranged on the tail.

Special Features: It appears that this species is often mistaken for *P. motoro*.

Potamotrygon orbignyi (CASTELNAU, 1855)
Synonym: *P. reticulatus, P. d'orbignyi, Trygon d'orbignyi, T. reticulatus, Toenura orbignyi, Taeniura orbignyi, Paratrygon reticulatus, Ellipesurus reticulatus, E. orbignyi.*
Habitat: The Amazon and Orinoco as well as in Columbia and Surinam.
Colour: The normal base colour is light-brown to dark-brown. A network of round lines makes a fine pattern, which can vary in intensity. Some specimens have small yellow spots on the edges. The skin is relatively smooth. The spines are missing from the tail.
Special Features: This species is often imported under the name *P. reticulatus.*

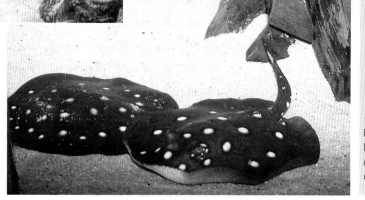

*Potamotrygon motoro in the Zoological Gardens in Frankfurt. The diameter of its disc measures approximately 80cm.
Photo: Hans Gonella*

*Potamotrygon leopoldi (P13).
Photo: Hans Gonella*

Potamotrygon schroederi (FERNÁNDEZ-YÉPEZ, 1957)
Synonym: n/a.
Habitat: Orinoco, Venezuela and Rio Negro to Manaus, Brazil.
Colour: This species has a light-brown to dark-brown-blue-brown base colour. Some specimens have white, yellow or black spots, which are arranged in rosette-like patterns. The spots decrease towards the edges.
Special Features: Apparently this species is imported occasionally. Despite this, nothing validated is known about caring for them in an aquarium.

Potamotrygon schuemacheri CASTEX, 1964
Synonym: Because of the different spellings of its species name that can be encountered, like schühmacheri and others, TANIUCHI clarified the description in 1982.
Habitat: Rio Colastiné Sur, Argentina.
Colour: These rays have a yellow-brown base colour with dark net-like patterns which decrease towards the edges. Dark spots can characterize the colour in the centre.
Special Features: Nothing validated is known about caring for them in an aquarium.

Potamotrygon scobina GARMAN, 1913
Synonym: n/a.
Habitat: The Amazon from Manaus to Belém and Rio Tocantins near Cametá as well as the Pará region.
Colour: The base colour is light-brown to rusty-brown. Many small white or yellow oval to kidney-shaped spots are concentrated especially around the edges.

In some specimens the spots form rosette-like or winding line patterns.
Special Features: These rays when young can be mistaken for *P. motoro*. Apparently they have been repeatedly imported into Europe under the name *P. motoro*.

Potamotrygon signata GARMAN, 1913
Synonym: *P. signatus, Paratrygon signatus.*
Habitat: It is possible that this species only lives in the waters of the Rio Paranaíba in Brazil.
Colour: These rays have a brown or light red shimmering base colour. Yellow spots can appear at the edges of the body disc especially in young rays. Otherwise they are normally lightly speckled. In some rays the spots form net-like or winding patterns. Mature stingrays have a double or triple row of spines.
Special Features: The young of this species can sometimes be mistaken for *P. motoro*.

Potamotrygon yepezi CASTEX & CASTELLO, 1970
Synonym: *P. hystrix, P. magdalenae, Trygon hystrix.*
Habitat: These rays are possibly only found in the waters around Maracaibo in Venezuela.
Colour: The base colour varies from light brown to pale grey/dark brown. The markings are usually made up of irregularly spread black, occasionally yellow, spots which appear more pronounced on specimens which have a lighter base colour.
Special Features: Nothing validated is known about caring for them in an aquarium.

Genus *Paratrygon:*

Paratrygon aiereba (MÜLLER & HENLE, 1841)
Synonym: *P. strongylopterus, P. aiereba* (not MÜLLER & HENLE), *Potamotrygon strongylopterus, Trygon strogylopterus, T. (Himantura) strogylopterus, T. (Paratryngon) aiereba, T. strongyloptera, Disceus strongylopterus, Ellipesurus* or *Elipesurus strongylopterus.*
Habitat: The Amazon and Pará region, north Bolivia, eastern Peru, the Orinoco in Venezuela.

with an irregular circular shape. At the front the body is slightly narrowed. Nothing is known about caring for them in an aquarium.

The "P-Rays"
As mentioned previously, the following list of species has no scientific validity in this form. Despite this, as the continually discovered unknown ray species have to be listed, the "P-numbering" system is intended as an aid until species identification can be clarified.

For example, in the *P. motoro* Peacock-

A pair of rays in an aquarium.
Photo: bede-Verlag

Colour: The base colour varies from light brown to grey-brown. The markings are a dark net-like pattern. Some specimens have light spots.
Special Features: This species is easily recognized; their body is relatively wide

eye Stingray group, individual specimens display different colours. Among other features they all share the typical "eye-pattern". Some of the forms described on the following pages could easily be assigned to a species, although the identification would still be doubtful without further detailed examination. For that reason, with a few exceptions, the "P-numbering" system is used.

P1, "Motoro"

Potamotrygon motoro from Peru, Brazil and Columbia.

The type includes rays which, in contrast to the well-known "Motoro forms" have smaller eyespots. In addition the centres of the eyespots are slightly orange. The base colour is an earthy mid-brown. The tail has a delicate pattern. Especially noticeable is the circular arrangement of the eyespots, which become smaller towards the edges.

Another type belongs to the Motoro variety P1, with eyespots that are arranged individually as well as next to one another. Many eyespots do not have a lightly coloured centre or it is very difficult to distinguish. The tail is lightly spotted.

P2, Motoro Variant

Potamotrygon motoro from Peru.

The eyespots of this type are evenly spread all over the body and they get smaller towards the edges. The eyespots are not always circular but can be either oval or of a different roundish shape. This type has more eyespots, but they are smaller. Smaller light spots lie between these eyespots. The base colour is dark brown and the tail is spotted.

P3, Motoro Variant

Potamotrygon motoro from Bolivia.

At first glance this type looks very similar to the P1 variant. The eyespots are a little smaller and are round but not entirely circular. They can be arranged next to one another and each one has a lightly coloured centre. They appear much fainter in the head region. The base colour of these rays is light brown, i.e. beige.

P4, Motoro Variant of the "Columbian rays"

Potamotrygon motoro from Columbia.

The body markings of the P4 type vary considerably from those of the other Motoro types. The black eyespots have a light centre which is usually circular especially in the larger spots and with a black centre. These outer edge of these circles in the eyespots can either be solid or interrupted. Some are a circle of spots. Light dots are evenly arranged between the eyespots. The base colour of these rays is mid-brown.

P5, Motoro Variant

Potamotrygon motoro from Peru

This type is easily identified by its near-black base colour. The black wide edges of the eyespots hardly contrast with the base colour. The centre of the completely circular eyespots is light brown to yellowish. The eyespots are arranged in a circular pattern on the body and diminish in size towards the edges. Light spots are dotted between the eyespots. The larger eyespots are surrounded by a circle of these light spots.

P6, currently an undescribed species

Freshwater stingray from the Rio Negro in Brazil.

The base colour of this type of ray is dark brown. Eyespots with a light border and a slightly darker centre are dotted around the edges. They are round but not circular. They are arranged evenly and become

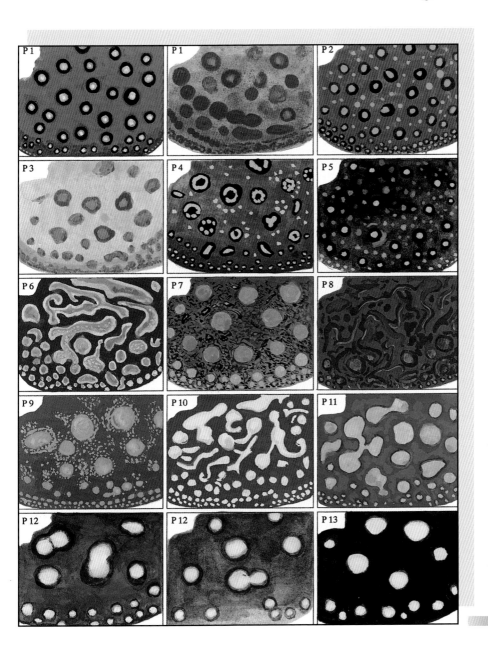

smaller towards the edges. The centre of the ray's body is decorated with large, evenly shaped, elongated brownish flecks with a light-brown border.

P7, currently an undescribed species
Freshwater stingray from Peru.

The dark olive-brown base colour of this type has an almost black pattern of many small spots and lines. Light olive-coloured eyespots with a more or less distinctive dark border lie in between. The arrangement of the eyespots can be compared to that of the Motoro variants.

P8, currently an undescribed species
Freshwater stingray from Brazil.

A type of ray with a dark brown base colour. The almost black pattern is similar to that of *P. orbygni* or *P. reticulatus* but much coarser. Some eyespots with dark borders and a mid-brown centre lie between the linear patterns especially around the edges. The eyespots are round as well as elongated.

P9, currently an undescribed species
Freshwater stingray from Brazil.

The base colour of this type of ray is dark brown. The fin edges have a border of several rows of light brown or yellow spots, which become smaller towards the edges. The centre of the ray's body has round spots of the same colour which are surrounded by many small light-brown-yellowish dots which gives the ray a very attractive colour.

P10, currently an undescribed species
Freshwater stingray from Brazil.

The markings of this type of ray are similar to those of the P9 type. The base colour is also dark brown, but the spots are yellowish and the small dots, which surround the spots of the P9 type, are missing. Individual spots are joined together to form a leopard-like pattern.

P11, currently an undescribed species
Freshwater stingray from Brazil.

The base colour of P11 is a very dark cobalt blue with evenly arranged light brown spots which diminish in size towards the edges.

P12, "Black Ray"
Potamotrygon henlei from Brazil.

A type of ray with a black base colour. The light yellow spots which are spread randomly either stand individually or touch one another. In some rays these yellowish or whitish spots appear blurred because of the fading colour. The fin edges are covered in lots of small light yellow spots, which surround the ray so to speak. The underside of some specimens of this variant can be black, especially the edges with yellow spots.

P13, "Eclipse Ray"
Potamotrygon leopoldi from Brazil.

A black ray with light strongly marked white spots which are spread more or less evenly across the body. Very rarely the spots are light yellow. In some specimens even the edge of the underside has a border of spots. The underside can also be partially black, otherwise it is white exactly like the P12 and P14 variety.

P14, "Leopoldi Variant"
Potamotrygon leopoldi from Brazil.

Some of the white or occasionally light yellow or beige spots of this type of ray are

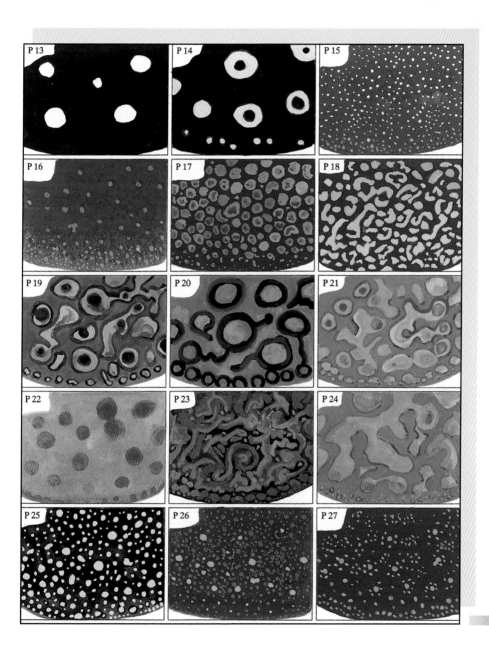

P 13 | P 14 | P 15
P 16 | P 17 | P 18
P 19 | P 20 | P 21
P 22 | P 23 | P 24
P 25 | P 26 | P 27

marked with a black dot in the centre.

P15, Antenna Rays
Plesiotrygon iwamae from Peru.

Rays with a greyish base colour and light whitish spots which are spread evenly across the body. Numerous small light spots lie in between, which reduce in size towards the edge. A variant of this has smaller spots which run right to the fin edges. The base colour is mid-brown.

P16, "Iwamae Variant"
Plesiotrygon iwamae presumably also from Peru.

The body pattern is identical to that of the P15 type, although the base colour is dark brown to almost black and the spots have a light brown colour.

P17, an unclassified species
Freshwater stingray, "Black-Tailed Antenna Ray" from Peru.

A type of ray with many light brown eyespots with a dark centre, which are spread evenly and reduce in size towards the edge. The multitude of irregular round spots on the dark brown to almost black base colour form a grid-like pattern.

P18, an unclassified species
Freshwater stingray, "Black-Tailed Antenna Ray" presumably from Peru.

This is probably the golden variety of P17. The rosette-like yellow spots together with the dark brown background create a leopard-like pattern.

P19, currently an undescribed species
Freshwater stingray, "Mantilla Ray" from Brazil.

Rays with a bluish-grey base colour.

Irregularly shaped light brown eyespots with black centres as well as elongated light brown flecks are spread all over the upper side.

P20, currently an undescribed species
Freshwater stingray, "Mantilla Ray" variant from Brazil.

Rays with a beige base colour and circular black rings which form a pattern. These rings decrease in size towards the edges.

P21, currently an undescribed species
Freshwater stingray, "Mantilla Ray" variant from Brazil.

A type of ray with a violet-grey base colour and beige eyespots with a light brown centre, which become smaller towards the edges. Many of the eyespots are joined together in the middle of the body, which creates a speckled pattern.

P22, currently an undescribed species
Freshwater stingray, "Orange Ray" from Peru.

The base colour of this type of ray is orange-brown with black spots that are unevenly arranged. Whitish-orange eyespots with a black border mark the tail.

P23, currently an undescribed species
Freshwater stingray from Brazil.

The coarse, linear dark pattern on a light brown base resembles that of *P. orbygni* or *P. reticulatus*. The fin margin is decorated with small light brown spots.

P24, currently an undescribed species
Freshwater stingray from Brazil.

These rays are marked with a light brown-grey, sandy base colour with yellowish-brown unevenly shaped spots. Some of the spots have a dark border.

P25, currently an undescribed species Freshwater stingray, "Otorongo" from Peru.

This type of ray has a dark base colour. The multitude of small yellow spots have a blurred border. In addition the fin margin is covered in smaller spots still.

P26, currently an undescribed species Freshwater stingray, "Otorongo" variant from Peru.

In contrast to P25 this type only has a few yellow spots just above the fin edges. Otherwise the delicate marking consists of

many small light brown spots which form a rosette-like pattern.

P27, currently an undescribed species Freshwater stingray, "Estrella" from Peru.

A type of ray with many small yellowish spots on a dark, almost black background. Some of the spots form a circular pattern each of which has another spot in the middle.

P28, currently an undescribed species Freshwater stingray, "Otorongo" variant from Peru.

This type is almost identical to P26. However, the base colour is mid-brown and the small light brown spots form a

Potamotrygon motoro.
Photo: Yvette Tavernier

delicate rosette-like pattern.

P29, currently an undescribed species
Freshwater stingray, "Hawaiian" from Peru.

With a dark brown base colour this type of ray also has rather delicate markings, created by numerous small spots which are arranged in a circular pattern. Some lighter spots on the body flanks complete the colour characteristics.

P30, currently an undescribed species
Freshwater stingray from Peru.

The spotted, dark body colour resembles that of *P. orbygni* or *P. reticulatus*. However, the light brown, evenly spread spots and the light dots at the fin edges are quite different.

P31, currently an undescribed species
Freshwater stingray, "Tigrinus; Tigrillo" from Peru.

This type of ray has contrasting markings. Lightly coloured lines create a brain-like pattern which contrasts very well with the dark brown background. The fin margins are covered with several rows of light brown spots.

P32, currently an undescribed species
Young of "Tigrinus; Tigrillo" from Peru.

The young ray described here has the same body markings as P31. The base colour is, however, beige and the markings are a very light brown. In addition the spots at the fin edges are light-yellow.

P33, currently an undescribed species
Freshwater stingray, "Estrella" variant from Peru.

This type has identical colours and markings as P27, only the arrangement of the spots creates a more delicate pattern.

P34, currently an undescribed species
Young ray of the "Estrella" variant from Peru.

This type has light to medium-sized spots on a dark background, the smaller dots in between – like P27 and P33 – are mostly missing.

P35, currently an undescribed species
Freshwater stingray, "Otorongo" variant from Peru.

Compared to P28 this type has slightly coarser markings on a dark grey background.

P36, currently an undescribed species
Freshwater stingray, "Motelo" from Peru.

A type of ray with light brown spots which form a rosette-like pattern and contrast nicely with the dark background. The fin edges are covered with several rows of light spots.

P37, currently an undescribed species
Freshwater stingray, " Belem Ray" from Brazil.

This type has a dark brown base colour with brown flecks and black spots, which get smaller towards the edges.

P38, unclassified species
Freshwater stingray of the "Belem Ray" group from Brazil.

This type is very similar to the *Potamotrygon henlei* species. The almost black background is covered with light spots with a black border. There are fewer spots in the centre of the body than near the edges. Some of the spots are not circular but kidney-shaped.

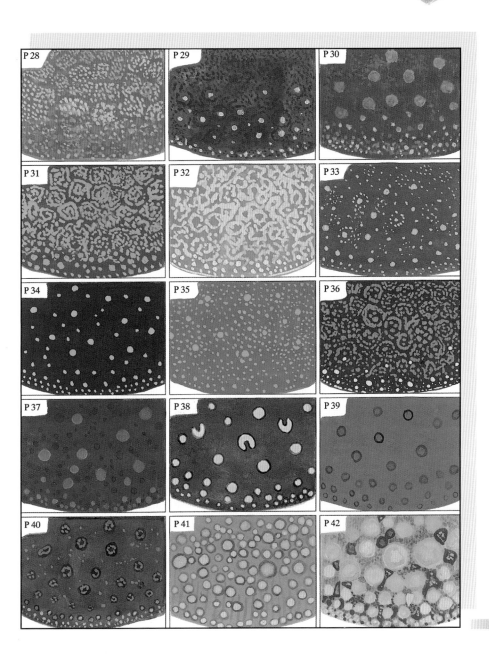

P39, currently an undescribed species
Freshwater stingray of the "Belem Ray"
group from Brazil.

A type of ray with a dark brown base
colour and evenly arranged brown
eyespots with a black border just like the
P. motoro species but differently coloured.

P40, currently an undescribed species
Freshwater stingray from Brazil.

This type of ray has a dark brown body
colour. Small light brown eyespots with a
dark border mark the fin edges. Black
round flecks are arranged evenly in the
centre of the body which often contain four
to five light brown spots arranged in a
circular formation.

P41, currently an undescribed species
Freshwater stingray from Peru.

Type of ray with a grey base colour and
many light brown eyespots. Numerous
small flecks appear between the evenly
arranged larger eyespots which appear
blurred towards the edges.

P42, currently an undescribed species
Freshwater stingray from Brazil.

This type has a beige base colour with
densely arranged light brown spots. Dark
brown eyespots in various shapes often lie
in between, which decrease in size at the
darker fin edges. The relatively strong tail
is very conspicuous.

P43, currently an undescribed species
Freshwater stingray, "Chocolate Ray" from
Brazil.

This type is slightly similar to *P. ocellata*.
The very dark base background is covered
with orange eyespots surrounded by many
small light brown spots. Numerous small

light brown spots are dotted in between.

P44, unclassified species
Freshwater stingray from Brazil.

The body colour of this type is similar
to that of the *P. motoro* species. The
eyespots are arranged more or less evenly
on the grey-olive background. They are
light brown with a dark border.

P45, currently an undescribed species
Freshwater stingray, "Sacha-Ray" from
Peru.

The light brown, sandy coloured body
of this type of ray is covered with ochre-
coloured irregularly shaped flecks with a
dark border. The outer fin edges are
covered with the same coloured eyespots
only they are smaller.

P46, currently an undescribed species
Freshwater stingray, "Mosaic Ray" from
Peru.

This ray has a particularly interesting
colour. Light brown flecks lie on the almost
black background, sometimes touching
one another, and they reduce in size
towards the edges. The edges of these
flecks are slightly darker. They are arranged
relatively closely to one another so that the
black background gives the impression of
a net-like pattern.

P47, currently an undescribed species
Freshwater stingray, "Florida Ray" from
Brazil.

The delicate yet distinct markings make
this type of ray something special. Light
brown eyespots with a border of several
dark dots are spread over a brown-olive
background. Small dark spots are scattered
in between and on the fin edges.

P48, *Potamotrygon scobina*
Freshwater stingray from Brazil.

This type of ray has a red-brown base colour. The yellowish, light brown eyespots are not very big, but are abundant. The relatively strong tail especially is covered with lots of smaller eyespots.

P49, currently an undescribed species
Freshwater stingray, "Tigre" from Peru.

The popular name "Tiger" does this type of ray credit. The tail is marked with black and golden-yellow stripes. The body has golden-yellow winding lines which lie close together on an almost black background.

P50, currently an undescribed species
Freshwater stingray, "Tigre" from Peru.

This type of ray is characteristic a young ray with pale colourings. The body markings are those of P49, but not the colours. They are light brown on a medium-brown background.

P51, currently an undescribed species
Freshwater stingray, "Tigre" from Peru.

This type of ray also characterizes a young ray with pale colours. The body markings and colours are more or less identical to P50. The gaps between the winding lines are a little smaller.

P52, currently an undescribed species
Freshwater stingray, "Tigre" from Peru.

This type of ray characterizes another young ray with pale colours. The body markings and colours are almost those of P50. The gaps between the winding lines are slightly larger.

P53, *Potamotrygon falkneri*
Freshwater stingray from Brazil.

The dark brown background of this type of ray has many smaller yellowish, mostly round eyespots with dark edges. The eyespots are close to each other, most of them touching, and decrease in size towards the edges.

P54, currently an undescribed species
Freshwater stingray, "Carpet Ray" from Peru.

Numerous densely packed, mostly touching spots – often six in number – each arranged in a circle on the dark brown, almost black, base colour create very attractive markings.

P55, currently an undescribed species
Freshwater stingray from Peru.

This ochre-olive-brown type of ray has individual flecks in various shapes with dark edges. Smaller eyespots with blurred contours are arranged around the fin edges.

P56, *Paratrygon aireba*
Freshwater stingray, "Ceja" from Peru.

This type of ray has the typical disc-type body shape, unique to the species. Thick light brown winding lines form the markings on the dark background. The very thin tail is conspicuous.

P57, presumably ***Paratrygon aereba***
Freshwater stingray, "Manzana Ray" from Brazil.

This type of ray also has the disc-type body shape that is typical of its species. The base colour is dark. The delicate light brown winding lines are made up of rows of dots. Light flecks lie between these rows. Another conspicuous feature is the very thin tail.

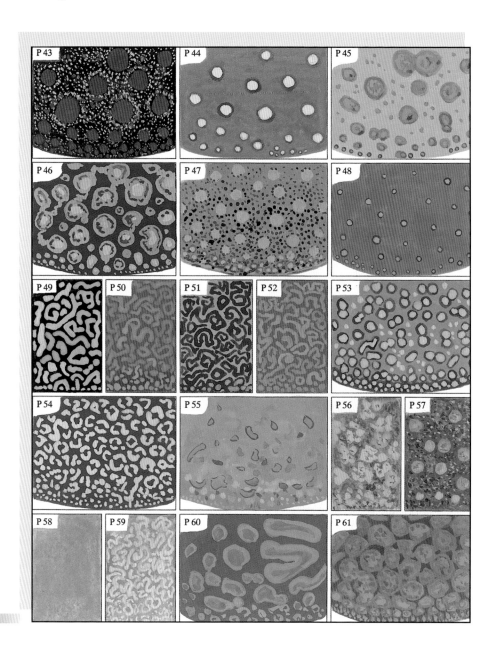

P58 currently an undescribed species Freshwater stingray, "China Ray" from Peru.

A type of ray with a short and very thin tail. The body colour is light brown all over.

P59 currently an undescribed species Freshwater stingray, "Coly Ray" from Peru.

This type of ray has a brown base colour. The light lines, arranged like brain windings, make a delicate pattern. Individual spots mark the fin edges. A short and very thin tail is also a feature of this ray.

P60, *"Potamotrygon hystrix"* Freshwater stingray from Columbia and Peru.

A type of ray with a dark base colour and light brown edged eyespots at the fin margins. The centre of the body is marked with large differently shaped spots of the same colour as the eyespots.

P61, unclassified species Freshwater stingray, "Columbian Ray" from Guyana and Columbia.

Type of ray with the characteristic body colour of *P. orbygni* or *P. reticulatus*, but with more contrast.

Freshwater stingray species from Asia and Africa

Occasionally species from outside South America appear on the market. These are not discussed in this book.

Apart from the South American potamotrygonids, the whiptailed stingrays of the Dasyatidae family are the second largest group of freshwater stingrays. The group comprises the genus *Dasyatis* with four species and the genus *Himantura* also with four species.

Their body size alone makes the rays with whip-like tails unsuitable for keeping in a domestic aquarium. Apart from special requirements in respect of water quality and size of tank, their care requirements are equally demanding.

Several other euryhaline species (i.e. able to live in waters with a wide range of salinities) are also known, which live at least part of their lives in freshwater, but they are completely unsuitable for home aquatics. Also some of the whiptailed stingrays only live some of the time in fresh water. Ultimately, keeping them in fresh water over long a period of time is impossible, which is another reason to avoid these species for home aquatics.

The protection of rays

Of the approximately 1000 cartilaginous fish species known today, only 43 species of ten genera and four families live permanently in fresh water. The members of the elasmobranch sub-class, i.e. sharks and rays, form the majority of the cartilaginous fish. Overfishing and the merciless hunt for sharks further destabilizes the already fragile balance of the marine fish world. And what about the situation regarding South American stingrays? This question cannot today be fully comprehended or answered. The fact remains that human population on the banks of rivers and lakes in South America is steadily growing. Despite road construction, the waters are increasingly used a transport routes. The effluent from human settlements is steadily increasing and often flows untreated into the rivers. Environmental toxins, such as agricultural

Potamotrygon
sp. (Tiger).
Photo: Yvette
Tavernier

pesticides and industrial waste, pollute the waters to an alarming extent. Landscapes are tragically changed through the clearing of large areas of ancient forests which causes erosion and the silting up of whole stretches of water. Exploitation of natural commodities poisons waters; for example, through large amounts of mercury used by gold prospectors running into rivers. Building gigantic dams to meet the ever-increasing demand for energy either destroys or changes life in and around water. At the same time overfishing of South American inland waters is increasing on such a scale that it is affecting even the remotest regions.

Considering the numerous threats to the environment, it is unlikely that the large stretches of water of the Amazon and the Orinoco will be spared. Rivers are in

would seem paradoxical to advocate taking fish stock for home aquatics. The evil has to be attacked at the root. Long-running projects to treat the rainforests and its stretches of water with great consideration – these can be compared with wild animal management schemes in Africa – seem to be a sensible solution, so that humankind, as well as the plant and animal world, can benefit from it. Controlled capture of ornamental fish is a justified part of these projects. This is in the interest of the aquatic trade, while the nature conservation authorities are also satisfied. One may hope that secure and lasting solutions are quickly found, otherwise this could adversely affect the nature of aquatics in the future.

South American freshwater stingrays should be paid more attention in many respects. On the one hand the sensitive ecosystems of the tropics and sub-tropics must be preserved through environmental protective measures so that rays with their specific requirements can find adequate habitats. On the other hand rays play an important role as "messengers" – namely to point out that the fishes kept in an aquarium come from an endangered environment. At least they draw our attention to the existing problems. Finally, demands are also made of every aquarist. Only if you are prepared to provide South American freshwater stingrays with exactly those conditions they demand in respect of care and reproduction should you devote yourself to keeping them. If not, then you should not get involved with this particular hobby, especially as the long-term care of rays does not allow for compromises.

many respects very susceptible to massive environmental changes, as is already the case with our seas. So one cannot ignore the fact that freshwater stingrays are especially unfavourably exposed to pressures from civilization. South American freshwater stingrays are an ecological and evolutionary treasure which must be preserved. Judging by the extent of the destruction of the environment, it

The ray aquarium

Although the ray aquarium does not vary much in its basic set-up from other types of tank, certain important considerations must be observed. The size of the aquarium alone and the necessary technology are a considerable financial investment. You also must be aware that looking after the fish will take up quite a lot of your time. With water changes, periodic cleaning of the filter and many other jobs, the keeper can face unexpected situations. It is quite possible that he will have to climb into the tank in order to adjust and correct any fittings. Keepers have to understand that rays are not "cute" aquarium fish but demanding wild animals. This means that aquarists must be extremely attentive when dealing with the sting-equipped ray.

Admittedly young rays that are available from aquatic shops can tempt one to an ill-considered purchase. But their bodies soon reach a diameter which will make the bottom dimensions of any ordinary aquarium appear inadequate. In actual fact South American freshwater stingrays, like all other members of this fish group, cannot be kept long-term in a normal aquarium with a capacity of 200 to 500 litres. Quite soon the aquarist will face the problem of what to do with a ray that has outgrown its home. In general large rays cannot be passed on to private fishkeepers as they simply do not have the facilities to provide sufficient space. Consequently, the sensible keeper will then invest in a larger aquarium. Better still, if he can't provide the necessary care, he should refrain from keeping freshwater stingrays.

Things to observe in the care of rays

Whether you are dealing with young rays or adults with a body diameter of 60-80cm, the ray's sting always presents a danger which should not be underestimated. The risk of the sting does not have to be feared excessively, but neither should it be considered harmless. This defensive weapon is used instinctively by South American freshwater stingrays – even when a careless movement in the tank is carried out with good intentions. If startled or pushed into a corner, rays do not attack with malice but simply to protect themselves from perceived danger. So no fishkeeper is immune from suddenly being

An undescribed ray species, Potamotrygon *sp.*
Photo: Yvette Tavernier

36

struck by a ray's sting. This danger is even greater in a tank which is too small as the rays hardly have any room to get out of the way. The effects of being stung by a ray vary. If the sting penetrates an arm, for example, it can cause considerable injury. However, the biggest danger is that the sting injects quite large amounts of poison into the wound. The effect of this poison can vary tremendously depending on the state of health of the injured person. Normally the wound will be very painful and swell up. Dizziness and sickness can last for hours or days, depending on the amount of poison which has been injected. With large amounts of poison the symptoms can last for weeks. Some injuries can possibly take between one to five months until completely healed. If not treated properly, wounds may become infected and they will be very slow to heal. Intense bouts of dizziness together with falling blood pressure, sweating, vomiting, muscle cramps or diarrhoea are all symptoms of large amounts of poison. Partial paralysis may even occur.

People particularly at risk are the elderly, people with a weak constitution, pregnant women and especially children. Even death cannot be discounted under certain circumstances. It is therefore imperative to consult a doctor if you have come into contact with the ray's sting. Even small scratches can cause unpleasant side effects which should be treated.

As a first-aid measure the wound must be thoroughly cleaned under cold running water. Then it must be cleaned again under hot water. This reduces the effects of certain poisons, which means some of the proteins contained in the poison are deactivated. Soaking the wound in hot water can also help. However, medical attention must be sought afterwards. The doctor should be made aware of the poisonous sting of the ray and informed about possible treatments. In general, European doctors have no experience in treating stingray injuries. It is best for the doctor to consult a Brazilian hospital about treatment methods. In any case the doctor should ensure the stability of the blood circulation. Painkillers or anti-allergens can be taken – after consultation with the doctor – which will ease the pain.

A tetanus injection should also be given if tetanus protection is not already effective. Occasionally, cortisone and antibiotics have also been used successfully

The purchase of an aquarium for stingrays involves a substantial investment and should be carefully considered. The ray aquarium must also have sufficient surface area, otherwise it is best to avoid keeping rays. Photo: Hans Gonella

Potamotrygon motoro *in the vivarium of the zoological gardens in Basel. This specimen is the result of a breeding programme at the Frankfurt Exotarium. Photo: Hans Gonella*

to help with the treatment. The intake of a Vitamin B complex can help speed up the healing process.

As an additional first-aid measure it also makes sense to buy an apparatus to suck the poison from the wound, such as a vacuum mini-pump. The pump is placed on the wound, and it sucks out the poison by creating a vacuum. This handy piece of equipment can be operated with just one hand, it does not need electricity and is not expensive. The Aspivenin mini-pump can be ordered online over the internet, or, if you are in France, may be purchased in pharmacies.

Having sucked the poison out of the wound, always follow the manufacturer's instructions; the wound should be cleaned as described earlier, disinfected and then medical attention sought. Incidentally, even discarded stings can cause injuries. Even after several years remains of dried poison can still cause poisoning if they come into contact with a wound. Stings kept as souvenirs should therefore be thoroughly cleaned and handled with care.

Nothing concrete has been established about the composition of the poison, but it is known to be potentially lethal. It consists, among other things, of semi-decomposed tissue in the sting channel. Consequently, the symptoms of massive blood poisoning can occur. One particular case is known from Europe where the hand of an aquarist had to be amputated as a consequence of incorrect treatment. In South America cases are known where people with weak hearts have died following a stingray strike. However, all this is not meant to create excessive fear of stingrays, but just to point out that South American freshwater stingrays must be handled with extreme care.

Aquarium size

It is imperative that the right size of the aquarium is chosen for the appropriate long-term care of rays. The larger the aquarium, the more comfortable the rays will be. While water volume is less significant, the surface area of the tank is a vital consideration. The minimum dimensions of the tank should at least be more than 1.8m in length and 1m in depth.

South American freshwater stingrays and arowanas are both fish which need a lot of space. If this is provided, they can be kept in a community tank. The rays prefer the lower and the arowanas prefer the upper water levels. Photo: Yvette Tavernier

For large ray species such as *P. motoro*, even tanks that are 4m long and 2m deep are almost too small. Equally important for the care of stingrays is the water level in the aquarium. Although rays tend to live on the bottom, the water level should never be too low as lively rays enjoy swimming in open water when they move equally through the upper and middle water levels. Therefore the water level must not be below 60cm – the higher the better. In addition tanks without a cover must have a rim of at least 20 to 30cm above water level to prevent the rays spilling out while they are swimming up and down.

When buying an aquarium it makes sense to acquire a strong stand. This ensures that the aquarium can be set up level. This also avoids having the aquarium at an angle if the floor is not level, which could lead to stress in the glass construction. It is also better for optical reasons not to place the aquarium directly onto the floor. Placed at a comfortable height of about 60cm, the space created below can be used to accommodate the filter and other equipment. Welded steel frame constructions make suitable stands. The sides of the stand can be clad with wood laminate panels. If they are fitted so that one of them can easily be removed, it will help to allow access to any equipment underneath.

Stingray aquariums with a capacity of 1000 litres and more are obviously tremendously heavy. It is therefore necessary to check the weight-bearing capacity of the floor. Also the aquarium should always be installed against a supporting wall where the weight-bearing capacity of the floor will be greatest. In rented accommodation it makes sense to have the static survey confirmed by an engineer. The landlord should also be informed about your intentions to set up an aquarium of this size. It is also advisable to check with your insurance company for adequate cover in case of water damage before setting up the tank; this ensures that in case of an accident you are adequately covered.

Position of the aquarium

The position of the aquarium in the living room will have a major effect on the way you look after it in the future. In order to avoid the problem of excessive algae growth caused by sunlight, the aquarium should be positioned as far away from a window as possible. It is much simpler to provide lighting in the tank which is also essential for aquatic plants. If the aquarium is only exposed to a few hours of sunshine in the morning and in the evening, then this should not cause any algae growth. On the contrary, natural sunlight is beneficial to plants and rays alike.

The harmonious integration of the aquarium into the living room is especially important to the aquarist. An aquarium has to fit in well to present a balanced picture. Access to the tank to carry out all the work must not be restricted. If the aquarium is open, you can also expect small amounts of water to spill out as a result of the strong movements of the rays swimming backwards and forwards.

The sitting room is a good place for an aquarium. From a settee or group of chairs which face the tank, one can enjoy watching what is going on. It is also easier to observe the aquarium and consequently rapidly spot any irregularities in the rays' behaviour.

By positioning foliage plants right next to the aquarium, you can create a fantastic impression of a tropical landscape. At the same time foliage plants can shade the aquarium from unwanted direct sunshine. If the aquarium has no cover, the space above it can also be used to position plants.

Aquarium equipment

There is a comprehensive range of aquarium equipment for fitting out a ray aquarium. At first glance, the choice of what is on offer in aquatic shops can be intimidating; and you ask yourself quite rightly which filter and what lighting to buy. In addition opinions on equipment vary depending on experience and preference, and one can expect different advice from different specialist shops. For the fishkeeper it is important that, for example, the filter systems fulfil their purpose and are easy to use. Not everyone finds every filter system equally easy to maintain. For the ray aquarium it is important to have efficient technology. The filter system especially has to be capable of removing any waste matter from the water. Murky water not only looks unattractive but is harmful to a ray's health.

Heating the water is equally important. The most important technical aids which are essential for a ray aquarium are introduced on the following pages. The systems are briefly described and, if necessary, a few notes added on their advantages and disadvantages. Product names are not mentioned on purpose in order to avoid voicing a preference for one particular product among the many that are available.

Lighting

The type of lighting is particularly important for the comfort of South American stingrays. Despite this general proviso, it is not advisable to over-expose the aquarium to light. Rays appreciate darker areas in the tank. In the wild they can escape to the darker, deeper water levels. But a certain amount of light is necessary for satisfactory plant growth, and although planting is less important in a ray aquarium than in other types of tank, you cannot do without it. After all, the plants have a beneficial effect on the whole aquarium environment. They also help to break down harmful substances in the water by using some of them as "food".

To guarantee satisfactory plant growth, two types of lighting are available. Firstly, there are fluorescent tubes which produce a lot of light with relatively low energy consumption. They are also available in different colours which, when combined, can create a variety of optical effects. Some light colours also enhance plant growth. Generally, for large aquariums, prefabricated hoods incorporating lights are not available from shops, so one is forced to buy a custom-made one. If necessary, one could combine several standard hoods for fluorescent tubes. If a hood is used which lies directly on top of the aquarium, a protective sheet of glass must be installed between the surface of the water and the lights. This is especially important to prevent the rays coming into contact with the lamps as it is quite common for rays to come to the water surface. An alternative to lighting hood would be to suspend fluorescent tubes in light fittings. These should be fitted approx. 50cm above the aquarium.

As another alternative mercury vapour or metal halide lamps are suitable for lighting a ray aquarium. These pendant lights are generally used for lighting open aquariums. Compared to fluorescent tubes they achieve equally good results for plant growth. They also supply sufficient lighting to low-growing plants when the water level is high.

For regular lighting it is best to install a timer. This ensures a certain period of light on a regular basis. Ten to twelve hours of light per day is recommended.

Water filtration

Adequate water filtration is one of the most elementary requirements of keeping rays successfully. Most problems with rays are caused because of poor water quality when waste matter and harmful substances are not properly removed from the water.

The aquarium filter has various functions. Firstly, it must ensure water circulation and at the same time spread the heated water evenly through the tank. Also it serves to remove suspended particles mechanically from the water. In addition the beneficial colonies of bacteria which

Metal halide lamps used as lighting.
Photo: bede-Verlag

live on the filter material break down some of the harmful substances that are found in the water. These mostly consist of fish faeces and left-over fragments of food and their presence must be reduced.

The volume of water in a ray aquarium

demands an efficient filter. Several filtration systems are available in specialist shops. They are basically divided into external and internal filters. Both types of filter have their advantages and disadvantages. In terms of their efficiency most types of filters can compete. However, they vary tremendously in respect of ease of installation and cleaning. It is therefore advisable to check out individual systems carefully before purchase.

External filters, as the name indicates, are fitted outside the aquarium. Hoses are used to effect water supply and drainage. They lead into the aquarium and have to be well fitted. External filters have the advantage of not taking up any space in the aquarium. They are generally installed underneath the tank but can also be placed next to the aquarium.

Internal filters are installed in one of the back corners of the tank. In order to prevent any unwanted movement of the internal filter, it should be glued to the wall of the tank with silicone. One advantage is that they are easily accessible and therefore easy to clean. For very large aquariums you may also consider combining several filtration system. For example an external filter, such as a canister filter, together with an internal filter could keep the water sufficiently clean. A so-called trickle filter could also be combined with one of the aforementioned filtration systems. Trickle filters are usually installed underneath the aquarium. Small amounts of water constantly flow over the filter medium and, provided enough oxygen is present, it increases the effectiveness of the cleansing bacteria which live on the filter medium.

Some filtration systems also have an integrated heater. These combined units have the advantage that the heater does not spoil the appearance of the aquarium. In addition powerful rays can either damage separate heating elements inside an aquarium or injure themselves on them.

When installing the filter it is important to show consideration of rays' typical behaviour. Every now and then they like to turn over the substrate. They pile up large heaps of sand and subsequently the filter inlets and outlets may get blocked with sand. If this goes unnoticed for only a short time, the filter will be damaged and will no longer work effectively. This in turn affects the water quality which is detrimental to the rays' health. So it is necessary to install the inlets and outlets in such a way that they cannot become blocked like this. The hoses leading to external filters must therefore be installed

sufficiently high up. Cleverly fitted stone constructions can prevent blockages of internal filters. Pipes or stone constructions should be fixed to the aquarium in such a way that the rays cannot move them. If necessary, they can be glued with silicone or other suitable fixtures to the glass.

Heating the aquarium

Freshwater stingrays live in tropical regions. For their well-being and to keep them healthy, they require appropriate water conditions. It is, therefore, essential to heat the aquarium. If the water is not kept at the right temperature, all the ray's vital physical functions will be affected. As a consequence they will become more susceptible to disease and may even die after a time.

The optimum temperature for South American freshwater stingrays is between 24° and 27°C. At night the temperature can fall by one or two degrees but otherwise the desired temperature should be kept constant as much as possible.

As mentioned in the previous section, combined thermofilters can be used for heating the aquarium. But of course the water can also be warmed by a heater which is fitted inside the tank. A heater should always be fitted close to the filtration outlet. This ensures that the heated water is spread evenly throughout the aquarium.

As with the filter inlet, it is important to position the heater in such a way that you avoid possible damage caused by any piled-up sand that may be excavated by burrowing rays. It should, therefore, be fitted well above the level of the bottom. A heater covered by sand is likely to overheat and break down.

An apparently undescribed ray species, Potamotrygon sp., in a quarantine tank at the premises of the importer, Aquarium Glaser.
Photo: Hans Gonella

Setting up and maintaining a ray aquarium

At first glance – apart from its size – the ray aquarium does not appear to differ much from any other home aquarium. But the installation and equipment of the tank require better planning and undertaking than is the case with conventionally set-up home aquariums.

The first difficulties can arise in transporting the aquarium if appropriate measures have not been taken. The weight of the aquarium requires the right sort of transport. Because of its size, it should be decided during the planning phase how it can be moved into the designated room. Before setting up the aquarium, you should decide on a suitable position in the room. Because of its great weight changes of position can hardly be carried out by one person at a later stage. When choosing the position you should consider the angle of the sunlight and where the nearest power socket is, so that you don't have to install an additional power supply at a later stage. The water supply should also be close to the aquarium to enable you to carry out regular water changes without too much trouble. The same applies to a drain. In general when carrying

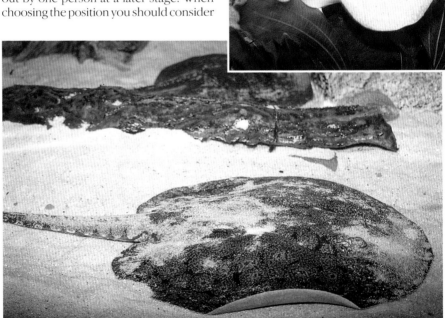

out a partial water change the water flows into a toilet or a low-lying drain. This is probably also the area where most of the fresh water for the aquarium will be taken from. To carry out a reasonable partial water change with a hose, the distance between the aquarium and water source should be no more than 20 metres. Additional equipment may be required to make it easier to handle the pipes over longer distances.

The purchase of an aquarium, as well as the choice of all the necessary technical equipment and fittings, requires careful planning. A hasty purchase often results in costly amendments later on. Therefore you should decide on how the aquarium is to be equipped and fitted and how all the maintenance can be carried out most efficiently. Easy access to the tank should also be considered so that the fun of your hobby is not spoiled by your having to move large pieces of furniture every time you carry out a water change or clean the aquarium.

Fitting out the aquarium

The fittings of the aquarium primarily have to be suitable for the rays. Large amounts of sand, stones and aquarium roots are needed. The set up and furnishing of the aquarium can easily take a whole day, if not longer, and only then if everything has been planned and prepared in stages. Bogwood should be soaked for at least a week in advance. This way it loses most of its water-clouding substances, which would otherwise soon turn the aquarium water into a murky underwater world. The substrate should also be rinsed to remove the finest suspended particles which would also cloud the water. Depending on the

substrate or type of sand chosen, the substrate material should also be cleaned more or less intensively. To clean the sand just superficially can take several hours, so this material should be prepared and made ready at least one day prior to fitting out the aquarium.

The types of materials used and the arrangement of the equipment must primarily fit in with the needs of the rays. The fittings in the aquarium must in no way inhibit the behaviour of the rays. This means that apart from observing the basic technical requirements, the appropriate fitting-out of the aquarium also makes a major contribution to the well-being of the fish. Feeling safe in their environment has a beneficial effect on their health. A properly fitted-out ray aquarium contains: sufficient swimming space, a few well-lit areas in the aquarium and, of course, a sandy substrate. The latter gives the rays the opportunity to bury themselves unhindered. That is why ornaments and plants should only be arranged either at the back or at the sides of the aquarium.

The decoration of the aquarium depends largely on the aesthetic tastes of the fishkeeper. If a background is to be fitted, this should be installed first. Afterwards the substrate is put in. Larger fittings, such as stones or roots, have to sit level on the floor of the aquarium otherwise the sand underneath them may move and cause them to tip over. Now the aquarium can be filled one-third full with water which makes it easier for planting. If it has not already been done earlier, technical equipment, such as the filter and heater, can now be installed – without connecting them to the power supply. Then you can fill the remaining two-thirds of the aquarium with water. In principle, the fitting out of the aquarium is now complete. But before turning the heater and filter on, just wipe the outside of the aquarium with a dry cloth to mop up any water splashes. The electrical equipment must not be switched on when it is wet, as this could be could be fatal to the aquarist. Immediately after switching on the power, you should do a final check to establish that the aquarium has no leaks and that all the technical equipment is working properly.

You should now wait for about two weeks before introducing the rays. During this time a perfect balance should be established in the aquarium environment. This just means that the bacteria on the filter start doing their job so that they can really carry out their cleaning and purifying task once the rays have been introduced. You can use some mulm from an already running filter on the filter medium to prime it. Alternatively the mulm or bacteria cultures can be bought from any specialist aquatic shop. But even if you don't prime the filter in this way, the bacteria will start naturally doing their job.

Once everything has stabilized in the aquarium, the rays can be introduced to their new home. You should do this by immersing the closed container in which the ray has been transported for about 30 minutes during which time the water temperature of the container will adjust to that in the aquarium. After that the ray should be released by carefully tipping the container so that the ray can swim into the tank. Additional rays should be introduced in the same manner. You should avoid introducing the rays with a net as their sting could get caught in it only too easily. It would also cause the ray unnecessary stress and the fishkeeper would have to cut up the net which would expose him to the danger of being struck by the sting.

During the first weeks the rays should be observed meticulously. Although the fish will be very shy at first, they will soon become inquisitive of their new surroundings. They should also not refuse food for longer than two days. It is not necessary to feed the rays on their first day in their new aquarium; it is better to offer them some food on the second day.

The rays must be able to bury themselves whenever they want to. This is the only way they will feel comfortable in an aquarium. However, this is only possible in a sandy substrate. If the substrate is too coarse, the rays will find it difficult to bury themselves.
Photo: Yvette Tavernier

Potamotrygon
sp. Coarse gravel
makes it difficult
for the ray to
bury itself in the
substrate.
*Photo: Yvette
Tavernier*

Substrate

The substrate is an important element in the ray aquarium. No other aspect of the tank is as important as this. Unless they are able to bury themselves, rays cannot express an important behavioural instinct.

So that they can bury themselves, the grain of the substrate must not be too large. Even fine-grained gravel is unsuitable for a ray aquarium. Coarse gravel makes it impossible for the rays to bury themselves. The ideal grain for the substrate is between 0.5 and 1mm. Very fine sand should also not be used, as it could get into the rays' eyes and gills which will have a detrimental effect on their health. The fishkeeper should also consider the composition of the sand. So-called sea sand which is used for marine aquariums must not be used in a ray aquarium. The sand is very chalky and will turn the water hard which is not suitable for rays which come from soft-water regions. Therefore you should only use appropriate chalk-free materials. Different types of sand in various colours are available from specialist shops, which are suitable for a ray aquarium. If none of the suitable types of sand is available, you can use the usual quartz sand with a grain of 0.4-0.8mm. The substrate must not be spread too thin either on the bottom. A layer of at least 10-20cm should be thick enough to allow the rays to bury themselves completely.

The sandy substrate enables the rays to carry out an important part of their behavioural repertoire. The sand is primarily camouflage for them. In the wild, once buried, the rays are well protected from any predators. In an aquarium the possibility of enjoying adequate camouflage give South American freshwater stingrays good living conditions. Lack of security in their environment, for example if there is too little or no sand at all in the tank, can be very stressful for rays. It is a well-known fact that stress is one of the main causes of illnesses, especially in fish. Not providing a suitable substrate for the rays for a prolonged period will considerably shorten their life expectancy.

Decorations

As long as decorations don't interfere with the quality of the water and the rays' natural behaviour, all the usual aquatic elements can be used to create a pleasing artificial environment. Ultimately, the fitting out of an aquarium is a matter of personal taste. However, the scale of the decorations should be in proportion with the size of the aquarium and should not restrict the rays' movements.

All "chalk-free and metal-free" types of rocks will create an attractive design. Multicoloured rocks especially can make an unusual picture. The most commonly used types of rocks are quartz, granite or gneiss, as well as porphyry. However, never use broken rocks with sharp edges as they can cause considerable injuries to the sometimes boisterously swimming rays, which then become an ideal target for infectious diseases. Large rocks, piled up high into "towers" also pose a similar risk. To give them the required stability, they

should be glued with silicone if necessary. However, this should be done at least one week before fitting out the aquarium. This will allow the silicone to set completely. Aquarium roots are particularly well suited to decorate the tank. Bogwood should always be chosen above other varieties as it has proved very suitable for aquatic purposes. No ray aquarium should be without some bogwood, especially if larger species of catfish are kept in the same tank. The pieces of wood provide hiding places for the catfish and also food as they regularly rasp them.

Artificial decorations are used increasingly to decorate aquariums. They

can imitate the look of a natural riverbank very closely.

Plants for the ray aquarium

Planting up a ray aquarium is quite different from adding plants to a normal home aquarium. You are limited in the choice of plants and their arrangement because of the prevailing conditions in the tank. You can only use robust plants and those which tolerate a low pH value. But not even plants with tough leaves are immune to being attacked by the rays. Occasionally the rays will eat them right down to the roots. Despite this, no ray aquarium should be without some plants. They have a beneficial effect on the aquatic environment and undeniably contribute to the overall look of the aquarium. Bogwood and the rich green of the plants complement one another in an attractive way. Only once the plants have been added does the aquarium truly reflect a South American biotope. It is for this reason that many aquarists try to cultivate their aquarium in a way that is as true to life as possible. However, this is not always a desirable aim. It is much better to choose those plants which will do well in a ray aquarium.

Occasionally the rays will go for the aquatic plants. Plants of the Echinodorus *species seem to be their favourites. Photo: Hans Gonella*

mean nothing to the rays, but they look attractive and are really there for aesthetic reasons. Apart from ceramic pieces, a vast number of synthetic ornaments are available from specialist shops. Some of the items look very natural and may have been painted. Man-made backgrounds have a certain artistic aspect which should not be discounted. As a ray aquarium should be fitted out sparsely in any case to allow for plenty of open swimming space, these backdrops can substantially improve the appearance of the aquarium. After all, the aquarium should be an asset to the living room. For example, with an appropriately printed background, you

Tip: Before you rush out to buy plants, give some thought to how the planting scheme will work with other ornaments in the aquarium.

To plan the planting scheme it is best to make a drawing of the tank layout, including decorations, other fittings, the chosen types of plant and their arrangement. At the same time think about how the positioning of the plants may be affected by the rays' habit of burying themselves. Admittedly, this is quite a problem when setting up a ray aquarium. Quite often the rays go looking for food in planted areas of the tank and subsequently dislodged plants will be found floating at the surface. One way to protect the plants from being dug up is by planting them in a separately built terrace at the back of the tank. These terraces can be approx. 15-20cm deep and tall, and they prevent the rays from digging up the sand in this area. Large rocks and ceramic pre-fabricated elements can be used for the construction of the terraces. The latter will have to be fixed to the floor with silicone to prevent them from being shifted by the rays. Another alternative is to set the plants in gravel-filled pots and secure them to the aquarium floor with large rocks. In general rays will ignore obstacles like this, but should they get in their way, only the pots will be moved around without the plants becoming dislodged.

Fixing certain plant species to bogwood is an interesting planting technique. Plants of the *Anubias* species, e.g. *Anubias barteri,* are suitable for this method. The Java fern, *Microsorum pteropus,* also thrives if grown on bogwood. Once established, the roots of the fern will attach themselves to the bogwood so that there is no further need to secure the plant. You can therefore use cotton to tie the plants to the wood initially. This will eventually rot and no longer spoil the appearance. Amazonian swordplants are also suitable specimens. They should be planted towards the back of the aquarium. By combining swordplants and large rocks,

you can create roomy spaces in the tank. Not only does it look attractive, it also creates separate zones in the aquarium. This allows the rays to occupy their own areas without having to make obvious contact with one another. However, these zones can only be created effectively in very large tanks. The giant hygrophila, *Hygrophila corymbosa,* or *Vallisneria americana* are also suitable for background planting.

As well as bog and aquatic plants, there are floating plants which can make a great contribution to the appearance of a ray aquarium. In addition they create those darker zones in the tank appreciated by rays. The water lettuce, *Pistia stratiotes,* and the water hyacinth, *Eichornia crassipes,* belong to this group of plants. A dense plantation of floating plants, which does not cover the whole surface, creates ideal conditions to keep surface-loving fish, such as hatchetfish, in the ray aquarium. They hardly ever come into contact with the rays so that they pose no problems. One thing to bear in mind with most surface fish is that they can easily jump high above the water surface, so for safety reasons the rim of the aquarium has to be quite high.

Aquarium water

The quality of the aquarium water has a direct effect on the rays. Poor water quality is to blame for most problems with keeping rays. So it makes sense to say a few words about the most important factors and this section is intended to encourage the fishkeeper to study the important subject of aquarium water.

"Clean" water is a liquid without any flavour or aroma. It consists of two parts

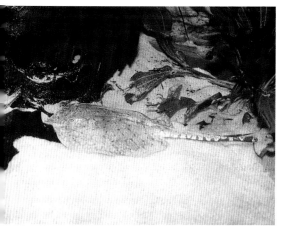

freshwater stingrays – can cope with astonishing fluctuation in water quality. It is a different story when the water chemistry is quite different to that found in their natural habitat or if the water contains excessive amounts of substances which are harmful to rays. The soluble substances found in water can be categorized into three groups. Firstly, salts; the inorganic substances are commonly known as minerals. They consist of two components once dissolved. The reason being that as soon as salt comes into contact with water, a water molecule will squeeze in between and split it into a positive ion (cation) and a negative ion (anion). These cations and anions can be detected in the water without establishing from which salt they are derived.

Secondly, organic substances, which are also found in water, are released in large quantities in aquariums. They are derived from food left-overs, the fish's faeces and decomposing plant matter. Gases are the third group. Oxygen, nitrogen and carbon dioxide play an important part in aquatics. There is a constant process of exchange between the gases in the atmosphere and in the water of the tank. This means that the gases present in the air and in the water are in a state of equilibrium which is dependent on the water temperature and the atmospheric pressure. However, this does not mean that the physical exchange process can adequately regulate gas levels in an aquarium. On the contrary. In general the rays absorb more oxygen from the water than can be exchanged from a still water surface. That's why the filter outlet has to be installed in such a

South American freshwater stingrays need soft water, with conditions which are similar to those in their natural habitat. Pictured here is the P. orbignyi or P. reticulatus species.
Photo: Hans Gonella

hydrogen and one part oxygen: H_2O. As water is an excellent solvent, it contains numerous dissolved substances. As fish (which of course include South American freshwater stingrays) are directly in contact with water, they have through their evolutionary development adapted to different types of water quality. If fish are denied water of the appropriate quality for a prolonged period, they will suffer sooner or later. Water is therefore not just an element in which the rays can move around, it is directly connected with all other life-maintaining elements in the aquarium.

To look at it uncritically, one could say that water is a fantastic self-functioning "chemical laboratory". Naturally the substances that water contains exist in a peculiar sort of harmony. Water not only contains millions of different substances but they also combine to create new forms. Other substances dissolve in water. So this poses the question: what effect does the variety of substances present in the water have on rays? However, this question cannot be answered completely and it is not of prime importance in respect of fish care as they have always adapted perfectly to changing conditions in their natural habitat. Therefore many fish species – and to a certain degree South American

way that the current disturbs the water surface which in turn improves the process of gas exchange.

The water values required for the care of South American freshwater stingrays must lie within a certain range. At the same time one has to remember that rays live in very soft waters in their natural habitat, so that the following recommendations should be observed. According to current findings most known freshwater stingrays can be successfully kept in water with a hardness of 4–12°dGH and approximately 4-5°kH, an electrical conductivity below 800μ/S as well as a pH of 5.8–7.2. The water temperature should be kept in the range between 24 and 27°C.

Water hardness

The hardness of water depends on the layers of rocks through which the water has percolated Depending on the composition of the geological layers of soil, more or less of the dissolved salts that cause hardness will be present. Chalky soil increases the water hardness, while in areas of granite, for example, water is more often than not soft. Depending on what region you live in, the quality of tap water can vary. The number of metallic ions present determine water hardness. They mainly consist of magnesium and calcium ions. Water hardness is separated into two categories, namely temporary and permanent hardness. Both together make up the total or general hardness which is measured in °dGH. Temporary hardness (also known as carbonate hardness) consists of calcium salts which precipitate out as calcium carbonate during the boiling process. They are measured as degrees of carbonate hardness in °kH. The pH value

is also influenced by the carbonic acid in the water and the carbonate hardness. The permanent hardness remains in solution after boiling. It consists of the salts from calcium and magnesium sulphates.

Total hardness and carbonate hardness can be measured easily with kits supplied by any good specialist shop. They also guarantee that the water quality does not fluctuate too much during a water change, which is particularly important when preparing the water. It is always advisable to keep your chosen water values as stable as possible. Constantly changing water conditions could have a detrimental effect on the rays' health.

If your tap water has a hardness of 10-12°dGH, it will be necessary to soften it. The carbonate hardness should not be significantly lower than 5°kH as this could destabilize the pH value. Your local water company can provide you with data on the water's composition and confirm if large amounts of disinfectants, such as chlorine, are added to tap water. If necessary, this can be removed from the water by using activated carbon. However, this is only necessary if the chlorine level is 0.2mg/l or higher, as chlorine escapes relatively quickly from the warm aquarium water.

pH Value

The pH value tells you whether your water is acid or alkaline. The abbreviation pH is derived from the latin "potentia Hydrogenii", which signifies the proportion of hydrogen ions (H^+) to hydroxyl ions (OH^-) in the sample of water. A range of 0 (extremely acid) to 14 (extremely alkaline) determines the acidity of a liquid; a pH value of 7 is a perfect balance of acidity and alkalinity. The pH scale is logarithmic, which means that a one-unit shift in pH registers a ten-fold change in hydrogen ion concentration.

Therefore two scale units indicate a 100-fold change in concentration. This is why most fish react badly to a sudden change in the water's pH value.

As indicated earlier the carbonate hardness effects the pH value of the water. In general the higher the carbonate hardness, the higher the pH value. The logical conclusion would then seem to be to achieve as low a carbonate hardness as possible. But this is only partly true. If the carbonate hardness is too low and the concentration of carbon dioxide suddenly increases in the water, it can lead to a so-called pH value crash which can be harmful to the fish. The carbonate hardness therefore has an important buffer function. Admittedly, this pH crash happens very rarely and in a properly functioning aquarium should not happen at all.

Ammonia-nitrite-nitrate

Large amounts of waste products are found in aquariums. Food left-overs, faeces and decomposing plants release proteins, which are converted into relatively harmless ammonium ions with the help of bacteria. Fish urine also contains ammonia. The ammonium ions also react according to the relative pH value. If the pH value is high, a chemical process starts, which turns the ammonium ions into toxic free ammonia. This process can however be avoided in a ray aquarium provided the water conditions are kept stable. If the pH value is low, ammonia is converted back into harmless ammonium. Through the action of aerobic bacteria, ammonia is converted into another toxic intermediate product, nitrite. Other bacteria then convert nitrite into the less harmful substance, nitrate. However, a severe lack of oxygen can lead to some of the nitrate being converted back into nitrite.

An apparently undescribed ray species, Potamotrygon *sp.*
Photo: Yvette Tavernier

53

Many freshwater fish can cope with surprisingly large amounts of nitrite, 80mg/l of nitrate is therefore no real problem. However, it is a completely different story with South American freshwater stingrays.

Based on different observations, it is believed that very high levels of ammonia, nitrite and nitrate can be harmful to rays. In particular, the skin damage that sometimes occurs in rays could be associated with poor water quality. If not rectified immediately, it could quickly prove fatal to the rays. Therefore, nitrate levels must not be higher than 5 to 40mg/l of water – the lower the better. A regular regime of partial water changes, a good biological filtration system and avoiding excessive feeding will help to keep nitrate levels in the water low.

Oxygen

An adequate supply of oxygen in the water is essential for the rays' health. The movement of the water surface caused by the filter outlet is sufficient to supply oxygen into the aquarium.

Problems with too low a level of oxygen will occur if essential hygiene, such as water changes and cleaning the filter, are neglected. Also high water temperatures in the summer can lower the oxygen level. But most significantly interruptions to or a breakdown of the filtration system can lead to acute lack of oxygen.

The first sign of lack of oxygen may be increased breathing frequency. It is also possible that at the same time the rays will absorb large amounts of oxygen into their intestines and subsequently float to the surface. This was observed when a filtration system broke down for several hours. Having cleaned the filters and got the system back into operation, the water conditions were restored within an hour and once the ray had expelled large quantities of air bubbles through its cloaca, it was able to sink back again to the aquarium floor.

Preparation of the aquarium water

When it comes to the water requirements for South American freshwater stingrays, it has to be said that these fascinating creatures are problem fish. Therefore, a certain amount of experience is necessary when preparing unsuitable tap water for use in an aquarium.

Of course it would be ideal only to keep those species of fish which are suited to your present water conditions, but only very rarely does sufficiently soft water come straight from the tap. Therefore, water preparation for aquatic purposes is unavoidable. To assist in the removal of the soluble salts present in tap water in the form of anions and cations, there are a variety of systems available from specialist shops. Firstly, there is a partial or complete desalination appliance. By using the complete desalination appliance, the water produced can be mixed with tap water until the required hardness is achieved. Other systems work on the principle of reverse osmosis. The aquarist has to choose which method will work best for him. However, the running costs of the various systems will probably influence his decision.

Carbon dioxide fertilization is another way of regulating the pH value by reducing the carbonate hardness. It certainly makes sense to use this procedure in aquariums with dense planting as the CO_2 supports plant growth. It is not necessary to add CO_2 to a ray aquarium that is only sparsely planted, although this method can be used to lower the pH level if the water is not too hard. It is however, advisable to monitor the water values continually with computerized measuring instruments. Fluctuations in these values should be avoided at all times. Additional literature on expert water preparation is available which gives in-depth information on this fairly complex subject. If you are insufficiently experienced in this area, it is strongly recommended that you undertake further research on water preparation before attempting to care for rays.

Regular care requirements

Your ongoing care regime is "make or break" for the successful care of rays. And although some of the subjects mentioned here might seem obvious to every aquarist, it makes sense to emphasize them again.

A Potamotrygon motoro with a very dark body colour.
Photo: Yvette Tavernier

Setting up and maintaining a ray aquarium

Neglect of these care requirements are the cause of the many problems.

Even the most efficient filtration system is no substitute for a partial water change. In order to remove excessive amounts of harmful substances, a partial water change carried out once every fortnight has proved to be very successful. A third of the aquarium water should be changed with fresh water. The precise amount and cycle period is less important than keeping the amount of changed water and the interval constant.

This helps the rays to adjust to the changing water quality in the tank. So it is also acceptable to change approximately a third of the water in the tank either every week or once every three weeks. While carrying out the water change, decomposing plant matter should also be removed. If necessary new plants can be introduced and any repairs or alterations to the fittings can be carried out. As any electrical equipment should be disconnected while carrying out these jobs, you must check that they are working properly once reconnected. The same applies to the water temperature which should be monitored at the same time.

The filter must be cleaned as soon as it no longer works efficiently. This can be carried out at any reasonable interval that will keep fluctuations in water quality to a minimum. Depending on the filtration system, it should cleaned every two weeks to every two months. Any longer intervals should be avoided as the rays produce large amounts of excreta. Cleaning the parts of the filter under a running tap must be carried out carefully, and this means removing completely any dirt which has collected on the filter under lukewarm running water. If the water is too hot, it will kill the bacteria living on the filter, which in turn will harm the biological efficiency of the filter.

Depending on the conditions in the aquarium, additional jobs will be necessary. For example, food left-overs must be removed after every feeding session. Occasionally the substrate should be cleaned of any dirt which will have collected on the floor of the aquarium with the help of a vacuum siphon.

In conclusion we would like to emphasize that all these jobs must be carried out with all necessary care and caution. Pushing the rays into a corner should be avoided at all cost. The next chapter "The behaviour of rays" give useful hints on how to keep risks when dealing with rays to a minimum.

A Motoro variant.
Photo: Hans Gonella

Plagued by algae – what now?

Excessive algae growth can become a real nuisance. Although algae belong in an aquarium, they should not be allowed to get out of control. In general excessive algae growth is an indication of an "unfavourable" care regime. Depending on the situation, excessive nutrients in the tank, the period of lighting and too much natural light falling onto the tank all can be the cause of unwanted algae growth. The fewer nutrients in the tank water, the sparser the algae growth will be. This can easily be achieved with regular water changes and an efficient filtration system. Healthy plant growth will also help to reduce algae growth as the taller plants absorb these nutrients. It is important to avoid disturbing plants once they have rooted if at all possible as this will interrupt their growth. Food can also be a cause of algae growth. Too much food or left-overs on the aquarium floor will provide additional nutrients for plants and algae. Large amounts of aquatic plant fertilizer can be another problem.

Apart from nutrients in the water, lighting duration also effects algae growth.

The time the lights are left on and their intensity determine which type of algae will grow in the tank. However, none of the lighting systems available from specialist shops is directly responsible for explosive algae growth. If used correctly, they are all suitable for aquariums. Algae growth occurs mostly in newly set-up aquariums. Even if a complete carpet of algae develops directly under the beam of the light, it is not necessarily caused by an incorrect lighting system. It is more likely that the complete "aquarium balance" has not yet been fully established. The algae should therefore be constantly removed and the lighting regulated. Given time – and this could take some months – the aquarist will surely succeed in restricting algae growth to a minimum. Lighting duration can be between 10-14 hours. This can be reduced by one to two hours if excessive green algae growth occurs. If brown algae develops, lighting should be increased by two hours. Beard algae and black brush algae are more difficult to control. They should be removed by hand. Blue algae must not appear in a ray aquarium. They are a sure indication of dirty water conditions. They like bacteria and develop in the substrate where they feed on decaying substances.

Fighting algae growth requires patience. It can take time, but remember that it is time well spent.

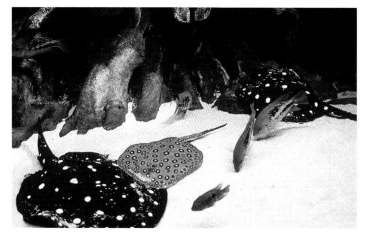

The behaviour of rays

species of ray in a tank. Keeping several species in a community aquarium has often led to problems. The small species of rays particularly are often attack-

Relatively little is know about the behaviour of South American freshwater stingrays. For example, next to nothing is known about the different reproductive behaviour of the individual species. Most behavioural observations come from watching aquariums. What happens in the wild has up to now escaped the inquisitive eye of the observer. Not least this is because the murky conditions of the waters in South America and their enormous volume make any satisfactory observations almost impossible. It is of course much easier to get a good insight of the rays' behaviour in an aquarium, although it is quite possible that they are subject to all sorts of disturbances in such a habitat. An unfavourable proportion of males to females, as well as the influence of other fish species, and of course the aquarium size can have an effect on various aspects of the rays' behaviour and consequently distort any behavioural observations. Insufficient space in the tank or keeping a type of ray that is not competitive in a community tank could make normal behaviour impossible or even lead to its premature death. The reason for the latter occurrence is often impossible for the aquarist to understand.

In order to avoid such unpleasant incidents, it is advisable to keep only one ed by their stronger and larger relatives. At feeding times especially it is possible that some smaller species will be attacked by the bigger, more powerful specimens and literally be torn apart. As these or other incidents don't always occur immediately the fish are introduced, but gradually develop or happen suddenly at a later date, it is often difficult to find an obvious reason for this. A certain amount of fishkeeping experience is required to draw conclusions that some noticeable behaviour is possibly the result of conditions in the tank. It is for this reason that freshwater stingrays are not recommended to the beginner.

As mentioned earlier, there are certain aspects of South American freshwater stingrays' behaviour which have yet to be explained. So it is important that the aquarist observes his fish with the utmost attention to help him understand their behavioural repertoire. The following comments are intended to help codify your own observations, or at least to add an extra strand to your own experiences. Admittedly, it is not always easy to interpret your own behavioural observations. It is especially difficult to form an opinion from a single observation. This behaviour may have a valid cause, but often observed behaviour may have happened just by chance. Therefore, your own behavioural

The behaviour of rays

observations require a high degree of "empathy" in order to avoid drawing the wrong conclusions and then sub sequently possibly harming the rays by taking certain measures, as a consequence of one mistaken belief. It is of course a completely different matter when certain unusual behaviours occur repeatedly. The first step is to establish a reason for this. If conspicuous behaviour leads you to believe that there is something wrong with the rays, or they could be suffering, it is important not to panic but to react moderately.

Behaviour in the wild

Observations made in the wild give us a clear picture as to which conditions have to be created in an aquarium to provide appropriate care for South American freshwater stingrays. Many questions cannot be fully answered as there is simply insufficient evidence available from observations in the wild. Some facts, however, are known and experience has shown that it is definitely possible to keep stingrays in a large domestic aquarium.

Most of the "snapshots" of stingrays seen in the wild were taken in shallow waters near riverbanks. Rays are found in shallow waters especially at night. During daylight one can recognize their resting places by the circular indentations in the ground. It is not known if they favour shallow waters particularly at night. It is assumed that they are looking for food. But another reason could be that deep waters have a lower oxygen level at night. This is because algae growing in deep water use most of the oxygen at night. But this assumption is purely speculative.

Opinions also vary as to whether rays are nocturnal or diurnal. But we assume that they are nocturnal. At least they become increasingly active at dusk and during the night. However, during the day we can watch them moving through deeper waters. In wide, deep waters they lie buried in the mud during the day and then suddenly emerge from their hiding place and swim away. Light is obviously not that intensive in deep waters from which we can conclude the lighting level in an aquarium should be equally low. Very often large groups of rays can be found in shallow lagoons, probably because these areas are rich in food and the rays can find plenty to feed on there. But it is not known whether rays develop any loyalty to particular territories. We can, however, assume that they cover large distances in search of food. Regarding the social behaviour of rays which gather together in large groups in one place, there is no clear indication that they form a close community.

Outside the mating season they most

probably live solitary lives but will happily tolerate the presence of other members of their own species and, as has been observed in aquariums, sometimes even make direct contact.

Behaviour in an aquarium

Many aspects of the behaviour of South American freshwater stingrays are very interesting. To what extent they are displayed in captivity depends primarily on the size of the aquarium. One of these, as mentioned previously, is the way rays bury themselves in the sand for protection. This type of camouflage was certainly one of the reasons for the successful evolution and continued existence of this group of fish. Hidden in the sand, rays can remain motionless for hours. If rays are denied this type of concealment, it will certainly cause them stress which could be harmful in the long run. And even if, under the aquarist's attention, the rays only make occasional use of this camouflage display, it represents an undeniable aspect of their lives. So it is imperative that a large sandy stretch is made available in the tank. At the same time you shouldn't be surprised if a ray constantly digs up the same group of plants or disturbs decorative rocks because it just likes to bury itself in that particular spot. In these circumstances you should let it have its own way and move the plants and rocks to another place in the tank. It is also noticeable that this practice of burying itself in the sand, as well as other behavioural habits, varies from species to species. For example, it would seem that the rays of the *Potamotrygon leopoldi* species bury themselves less often than the Peacock-eye Stingrays of the *P. motoro* species.

> **Tip:** With appropriate care the (initially) very shy South American freshwater stingrays soon become very trusting in their new environment.

With great curiosity, which is almost proverbial, these rays explore every corner of the aquarium. They may constantly blow into the sand with remarkable persistence in order to find food. While doing this they will even search the sand under rocks. During their explorations in between rock structures they also use their ventral fins, although there is no reason why they should be doing this. They particularly like blowing into the cavities of stones with holes, where they are presumably hoping to find some food.

Apart from the extensive daily food searches, an enormous urge for discovery and play has been noticed in rays, during which they can literally dig up the whole aquarium floor and pile up the sand in large heaps. They also notice even the smallest change in the tank. Just the addition of a new river stone can awaken their curiosity and they will thoroughly examine it. Even moving the stones in the aquarium can provide a welcome change for the rays. However, making big changes to the tank should be avoided as the rays can get very agitated if their habitat is suddenly changed. Rays which are kept in an unfurnished aquarium mostly display a phlegmatic temperament.

Despite their bottom-orientated way of life, rays have a strong urge to be active to which they want to give rein. So they swim with strong fin movements through the middle and upper regions of the aquarium

61

water. When doing this they can occasionally reach such great speed that, as they swim upwards along the aquarium wall, they slide over the rim and land out on the floor. If such an accident remains unnoticed, their gills will dry out quickly and the ray will die. But even if the rim of aquarium is high enough to keep them in, they will still leave behind signs of their swimming escapades by splashing water onto the floor as they turn over at the surface of the tank.

It can get really hectic in the aquarium when two rays get in one another's way. This is usually a sign of lack of space. However, arguments between two males occasionally happen. Some males may also display very aggressive behaviour, although rarely towards females. In extreme cases these clashes can result in the death of the weaker fish. However, arguments are rare between same-sized fish of the same species so that freshwater stingrays can generally be described as peaceful fish. When it comes to food though, the more robust species will assert themselves over the weaker species. Hence it is advisable only to keep same-sized fish of the same species in a community tank with one male and one or more females.

When two rays fight, it is quite possible that they will inflict injury with their stings. If this is the case, the sting will not necessarily cause great harm. As long as no important body parts are affected, the poison will simply cause a clearly visible round skin coloration around the entry wound. The wound and the affected skin tissue heal quickly without any harmful after-effects. Although this description may imply that they are solitary creatures, rays are in fact extremely sociable fish. They lead a predominantly harmonious life with their relatives even to the extent that they sit on top of one another to enjoy close body contact. Even during feeding they hardly ever get in each other's way and should obstructions occur, they simply nudge one another to get to their chosen food. Sometimes during mating female rays can suffer minor injuries to their fins that are inflicted by the males. However, they are no cause for concern and a part of the normal mating pattern.

Given time the rays can become very trusting towards their keeper.

Their well-developed love of play and their natural curiosity actually encourages hand-feeding. With growing trust some rays will even let themselves be touched. This, however, poses a potential risk which should not be underestimated. A careless movement or withdrawing a hand too quickly can have painful consequences. As freshwater stingrays are wild animals,

decide to take a closer look at the rays which could easily cause them to put their hands in the water. So it is sensible, when necessary, to restrict access to the aquarium. For example, a cover with a lock can be fitted.

South American freshwater stingrays are extremely agile. Don't be deceived by their often calm behaviour; they can react as quickly as lightning and sting you by lashing out with their tail with unbelievable speed. If startled while resting, they'll move like a flash, and the aquarist may make some thoughtless movement which the ray will interpret as a threat. Rays are definitively not vicious, but they will use their defensive weapon if they feel the need to do so. And they cannot distinguish between a well-meant intervention, a display of affection or an attack. As soon as they are confronted with an unusual situation, they will instinctively try to defend themselves.

Knowledge of rays' behaviour and fishkeeping experience can only be acquired over time. Only an intense preoccupation with these fascinating fish will allow the aquarist to start assessing rays correctly. For this he has to be alert and attentive in his care routine.

From time to time rays can become very lively as can be seen in this picture of a pair of P. motoro. Photo: Yvette Tavernier

responsible aquarists must not allow themselves to be lulled into a false sense of security and put themselves in danger. It is advisable to avoid direct contact with the rays while carrying out maintenance work or feeding, and a plant cane or a similar instrument should be used for working inside the aquarium.

Important precautionary measures

When talking about the behaviour of rays and how to look after them, the potential dangers cannot be over-emphasized. Even if the warnings in the following section begin to sound repetitive and boring, it will certainly help you to learn how to judge rays. We have already warned about rushed movements and the hazards of touching rays. Other family members and anyone looking after the rays while you are away must be made aware of how poisonous rays are. Children especially must be protected from them. They may suddenly

Reproductive behaviour

Just like all other fish, South American freshwater stingrays have developed mating rituals which are unique to their species. Precise knowledge about their mating behaviour through the studies and interpretations of behavioural scientists does not exist. Nevertheless some behavioural characteristics give an insight into their form of mating. However, this is only very rarely observed in an aquarium.

This is not so much because the fish tend not to mate in a tank, but because mating usually takes place under the cover of darkness. Traces of their mating can then be seen on the female in the morning. Her fin edges are often damaged as a result of the intense courtship of the male. Every now and then whole chunks are missing from the fin margins which gives the impression that a considerable fight had taken place. This heavy "game" can be described as a courtship display.

Initially the consenting partners will swim next to or above one another more often than is normal. They will constantly encircle one another and at the same time try to grab hold of each other at the back next to the base of their tails. It is predominantly the male who tries to persuade the female to mate by hanging on to her fin margins. During the second phase of courtship the female indicates that she is interested in mating by responding to the male's encircling and tugging and allowing him to move closer.

During the third phase the male is able to slide sideways underneath the female and, with their stomachs touching, lift her up. In this position the male can insert one of his secondary reproductive organs into the cloaca or female sexual orifice. The male reproductive organ guarantees that his semen are inserted into the female's body without great loss which helps to ensure successful fertilization of the eggs. The mating ritual can last between 20-30 minutes, but very little is known about the precise length of time. It is quite possible that mating lasts for several hours. Based on certain observations it is safe to assume that mating takes place several times over the following nights. Virtually nothing is known about the stimulus which leads to mating. However, other females in the tank will be left alone by the male from which we can conclude that the receptive female sends some signals to the male indicating her willingness to mate. After mating the male will leave the female alone and he does not show interest in her again until the next mating season.

Rays and their companion fish

The lively behaviour of stingrays raises the question of whether it is possible to keep them in a community tank with other fish species, and not least because they rummage in and around the substrate while looking for food. The answer to this question of compatibility with other species is: yes – keeping rays with other suitable fish species is advisable. One might go as far as saying that a multi-species community can have a positive effect in the aquarium and will create a more interesting environment for the rays and other fish.

Admittedly problems between rays and other fish species cannot be eliminated completely. Small and slim fish species which live near the bottom or small fish moving around the substrate at night could possibly be considered a welcome delicacy by the rays. This is especially the case if these fish are unable to hide from the night-raiding rays. In addition certain territorial cichlids may have confrontations with the rays and these are usually fatal for the cichlids. During the mating season some cichlids want to inhabit the same areas in the aquarium which the rays use as resting places.

Discus are an example of good companion fish for rays. However, an aquarium environment that lacks

stimulation or enough space may be problematic for the fish. If a discus should find itself in a confrontation, the ray may kill it without even using its sting. It is quite possible that the physically superior ray will literally bury the discus underneath itself. The discus will try to defend itself with strong and panicked movements – but without success. This is however not the normal behaviour of a ray as the discus is not naturally a ray's prey. Only unfavourable conditions in the aquarium can be blamed for this. A ray would not even attempt to eat a discus. Apart from discus, Chocolate Cichlids, *Hypselecara*

temporalis, even dwarf cichlids and many other rock-loving cichlid varieties can live happily with rays, provided all species have the same water quality requirements.

Also among the tetras there are many species which can share a tank with rays, such as the Buenos Aires Tetra, *Hemigrammus caudovittatus,* and other small members of this group of fish. Plenty of hiding places, which are used for resting and sleeping, are a basic requirement though. Large suckermouth armoured catfish species are ideal as companions. They require roots which they can also use as hiding places.

Above left: Small fish species kept in a ray aquarium must be able to hide at night between plants and roots.

Above: A community tank from Ottrott in France. Rays, discus and an eel share the aquarium. Photos: Hans Gonella

Feeding rays

South American freshwater stingrays are so-called detritus-feeders, which means that, among other things, they feed on small organisms and larger living creatures that live in and above the muddy waterbed. But they also feed on large quantities of plants. These slowly decaying plant and animal remains form the upper layer of the riverbed and are therefore an important part of the rays' diet.

Their body shape and way of life alone are an indication that they are specialist feeders. They dig up all sorts of living organisms from the muddy waterbeds. Apart from insect larvae, they also feed on worms and crustaceans. They can even crack the hardest shells effortlessly with their strong teeth. Small fish resting on the water bed are welcome food as well. Despite this characteristic, stingrays cannot really be categorized as predatory fish.

Very little is know about their diet in the wild. However, this does not mean that they don't require a balanced diet in an aquarium, especially as ample information on their feeding habits is available from experienced aquarists. A specific and balanced diet in moderate amounts provides all the essential nutrients. Adult rays simply need to be fed their basic food twice or three times a day. This can be supplemented with a variety of titbits two or three times a week. It is important that the diet is varied in order to avoid the rays refusing their food.

South American freshwater stingrays generally eat substantial amounts at any one time. It is easy to tell by their behaviour if they are hungry. They become restless and constantly hover above the substrate while blowing into the sand. As their hunger increases, they will even swim up and down the aquarium walls. On the other hand, if rays refuse to eat over a long period, it is usually an indication of health problems.

Tip: You can tell if a ray is well nourished by the "fat deposits" which appear as bulges either between or behind the eyes on the upper body.

Healthy rays can reach a considerable body weight of 3-5kg.

Through regular and moderate feeding you will prevent health problems. Too much or too little food can lead to diseases and subsequently reduce the life expectancy of rays. Therefore it is important to monitor their state of nourishment. Excessive and over-fatty foods, as well as food rich in carbohydrates, can also cause organ damage. However, undernourished rays with sunken body parts are unlikely to survive.

Choosing the right food

The most common food for rays are peeled shrimps and fish meat. Both are an excellent food as long as they are not fed exclusively. Rays are particularly fond of unpeeled shrimps and they make use of their flat, grinding teeth while eating them in order to crush the shrimps' shells. Pieces and strips of fresh- and saltwater fish can be fed but freshwater fish is the preferred food. Apart from trout, fillets of whitefish and river perch are also suitable as food. The meat of the latter two has a dense consistency which helps to prevent cloudiness in the water at feeding time. Mussels and small pieces of squid should

as well as ten per cent pre-cooked brown rice or ground corn. The remaining five per cent are made up of aspic gelatine which is added to bind the mixture together. Ten litres of water are mixed with 0.6 litres of gelatine. Added to this mix are raw eggs, two to three level tablespoons of a

While searching for food, rays blow into the sand in order to flush out creatures such as insect larvae. Photo: Hans Hollenstein

also be fed occasionally. Live food is also recommended. Earthworms, red mosquito larvae or small shrimps, such as brine shrimps, are also a favourite of adult rays. Live food has the advantage that it awakens the rays' motivation to hunt for food themselves, and rummaging through the substrate also keeps them occupied.

All these foods have a high content of animal protein and also a relatively high fat content. Supplementing this type of food with plant food will prevent digestive problems. It is therefore important to add a variety of plant foods to the aquarium on a regular basis. For example, cucumber and salad leaves are amongst the rays' favourites. It goes without saying that all types of food, both fish and plants, have to be fresh and washed in order to keep the rays healthy.

A food mix which you can prepare yourself has been proven to be particularly suitable as it will provide all necessary nutrients. The recipe comes from the Frankfurt Zoo and is made up as follows: the fresh ingredients are 30 per cent beef heart, 20 per cent freshwater fish, 15 per cent squid and ten per cent beef liver plus ten per cent carrots, apples or other fruit,

vitamin-mineral mix as well as ten ground vitamin capsules and approximately a quarter-teaspoon of canthaxanthin per ten litres of mixed food. If, for nutritional reasons, the meat content is too high, this can be replaced with fish meat. Canthaxanthin is also not essential. All ingredients can be mixed in a blender and the eggs added afterwards. Then the mix is heated in a saucepan to around 60°C. To prevent the mix sticking to the bottom of the saucepan, add a little water, but don't make the mix too runny.

Once it has reached the right temperature, the aspic gelatine is added slowly and stirred thoroughly so that it can completely dissolve. Vitamin supplements are added last. Then pour the mixture into a shallow dish and allow it to cool as quickly as possible. The quicker the gelatine mix cools, the firmer its consistency will be. It is best to place it in the fridge. When it has cooled down, remove the mixture from the dish and cut it into small cubes, which can then be fed in small portions to the stingrays while the rest is kept in the fridge. This staple food can be fed two to three times a week and may be supplemented with a variety of

67

titbits. Young rays, however, can be fed several times a day with the base mix to ensure healthy growth.

How to feed properly

There is actually not much to say about this topic. However, certain aspects have to be considered in order to avoid unwanted surprises. In the wild freshwater stingrays find their food predominantly in the muddy water beds. The same behaviour is demonstrated in the sandy substrate of the aquarium. The reason being that rays also search for food on sandbanks in South American waters. In their constant search for food they blow into the sand, but they are also happy eating any food which lies on the ground.

Feeding problems usually occur during the first weeks in the aquarium. Very often the rays will swallow their food but, having chewed it for a little while, will spit it out again. This makes it very difficult for the aquarist to tell if the rays have eaten anything at all. This type of behaviour is a sign of food refusal, as healthy rays usually do not refuse appetizing food. Refusal of food can occur for a number of reasons. Certainly weak and ill rays refuse food. But a change in their environment can also lead to a loss in appetite. And lastly, the new unknown food or portions that are too large can be a problem for the rays. Live

food is ideal to encourage rays to eat. They can be reared on red mosquito larvae or earthworms from unfertilized meadows, which have not been sprayed with pesticides. You can also feed very small amounts of finely cut shrimps or fish at the same time. To ensure that all the food is eaten, do not place more than two to six pieces into the aquarium. Only when they have definitely been consumed should more food be added. Then gradually increase the amount of food. By doing this you will get a good idea of how much food your rays actually need.

It is important for young rays to be fed regularly and with a variety of foods. Several small feeds each day also promote healthy growth. Older rays on the other hand can benefit from two "fasting days" per week. For rays to refuse food for a couple of days is nothing unusual. These self-imposed fasts are really nothing to worry about.

After each feeding session any uneaten food should be removed from the aquarium after a couple of hours. This will prevent the food from decaying which could subsequently lead to poor water quality. The same procedure applies to live food. Red mosquito larvae should only be fed in quantities which the rays will eat at once. If left uneaten in the substrate, they will begin to rot and spoil the water quality.

Breeding freshwater stingrays

For a long time the mating habits and the embryonic development time of South American freshwater stingrays were a mystery. Even today certain aspects remain unclear. In particular, mating frequency and the hatching of the young do not reveal a precise time pattern. The young of not-fully-mature pregnant females seem to be born over a period of several days or weeks. Quite often it is safe to assume that mating took place repeatedly, which can be seen by the marks on the female's fin margins. Births can also take place at different intervals. Young rays can be continually born over intervals of between two months up to more than a year. It was therefore assumed for a long time that embryonic development must take approximately five to ten months, especially as the gestation period of other cartilaginous fish was found to be in excess of a year.

Based on observations made in the Exotarium of the Frankfurt Zoo we now know that embryonic development takes two and a half to three months. It was also established that the female's fertility changed with increased body size. While young females can only give birth to one or two young, larger mature females can give birth to six to eight or more young. At the same time the young of the smaller litters are substantially larger than of the larger litters.

Even today contradictory information on embryonic development is published in specialist publications. This is a sign of how little is known about the natural reproductive biology of South American freshwater stingrays. One thing is certain though, South American freshwater stingrays are ovoviviparous (the eggs hatch out within the mother's uterus) and therefore belong to the livebearing fish. Each embryo is surrounded by an egg membrane. The embryos feed on a relatively large supply of egg yolk. Once the embryonic development is complete,

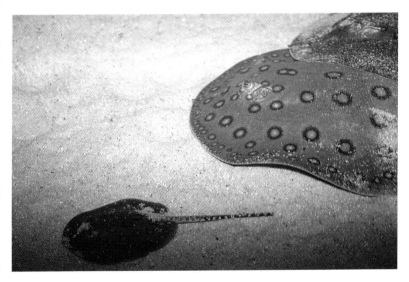

This picture shows a mother, father and newly-born daughter ray living harmoniously together. Photo: Hans Gonella

the young rays hatch from this membrane before they are born.

The following information does not give complete breeding instructions due to the current lack of sound scientific knowledge. However, using our existing knowledge it is quite possible to encourage South

American freshwater stingrays to breed in domestic aquariums, as several offspring have been produced in home aquariums, a fact which cannot be dismissed as mere coincidence.

Requirements for breeding

Certain requirements will have to be observed in order to make successful breeding of South American freshwater stingrays a possibility. Personal experience of keeping tropical fish is certainly an advantage in helping you to provide constant conditions in the aquarium that will make breeding possible. The aquarium has to be sufficiently large to accommodate a pair of rays. A tank with an approximate water capacity of 2000 litres just about meets the minimum requirement. However, the total water volume is not as important as the floor area which the fish will use. Ultimately though, stable living conditions will make the difference between success or failure. Among all other factors, such as the harmony that exists between male and female and food supply, water quality plays the most important part in breeding stingrays. It might also be sensible while trying to breed them in a domestic aquarium to keep the rays separate from other fish species; this ensures that their normal living conditions are not unduly disturbed.

When choosing a pair of rays for breeding, the male and female should be of a certain size. With a body diameter of around 40cm, it should be obvious if the pair are compatible. Occasionally the male may attack the female without any reason and cause her serious injury.

Peacock-eye Stingrays, *P. motoro*, reach their sexual maturity after approximately two years. This should more or less hold true for other ray species. Sex is determined in the same way for all species.

> **Tip: Males have claspers, modified pelvic fins, on each side underneath the tail, which cannot be missed.**

One clasper is used per mating session, the second one being a "reserve". The females have no such secondary sexual organs. South American freshwater stingrays remain fertile well into old age. With appropriate care they can live from 15 up to 25 years, and perhaps even longer. To what extent different species interbreed and produce young is not known. If they are closely related, it is quite possible. It makes sense therefore to keep just one species in an aquarium. At the same time you should only keep one male per female in order to avoid rivalry developing among the males. It would, however, be possible to keep one male with several females in an exceptionally large aquarium

Water for breeding

The importance of the appropriate water quality has already been mentioned in this book. South American freshwater stingrays are quite tricky in this respect. As all aquarium-kept species known to date originate from regions with very soft water, the water quality in the tank has to be the same as that in the wild. This is even more important for breeding purposes. If you are keeping rays just as a hobby without any intention of breeding, the water values can be marginally higher. Water used for breeding must have a pH value of below

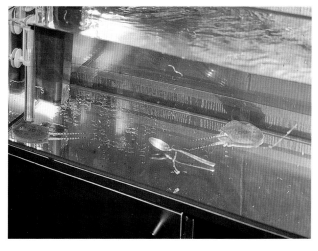

6.5 and the overall hardness must not exceed 5° dGH. Carbonate hardness has to be equally low. Water used for partial water changes must not contain any chlorine, which could possibly be harmful to the rays' health. The same applies to any possible residue of medication which may have been used previously to treat diseases. If necessary, the breeding water should be filtered through a charcoal filter. No significant amounts of nitrate and nitrite must be present either, as these would foil any breeding attempt. It is therefore imperative to test water chemistry regularly so as to establish that the water is free of harmful substances.

Rearing young rays
Before they are born, the movements of the embryos can be seen in the mother's body. Another indication of pregnancy are the curves on the female's upper side. The baby rays hatch from the egg membrane shortly before the female gives birth. They emerge from their mother's body back to front with the tail lying on top of their body, which prevents the mother from being injured by their soft, but already well-developed, sting. The sex of the new-born rays can also be determined immediately. The males have the same modified pelvic fins as their father only they are proportionately smaller. These claspers will later serve as secondary sexual organs.

Rays bred in captivity often give birth prematurely or the young are still-born. Quite often the fully developed and healthy looking young die within a couple of weeks. It is also quite common for them to refuse food. What could be the reason for this? Apart from inappropriate care, miscarriages or the deaths of baby rays

shortly after birth are probably an indication that the female is too young and the embryos could not fully develop in her body which is too small and lacking in hormones. This assumption is only based on the observation that the same female can produce larger and more resilient young later in her life.

After birth, the young can either be kept in the same tank or be moved to a nursery tank. If space is no problem, it is unlikely that the parents will attack the young. Even if an adult literally covers a young by settling on top of it, the young ray knows how to draw attention to itself by violently wriggling around – normally the adult will quickly swim away. However, a home aquarium is often short of space and in such cases it is best to move the young into a separate tank. It is essential that the water conditions in the nursery tank are exactly the same as in the tank where the young rays were born, otherwise they will die quickly. The newly born rays are not

A baby ray, P. motoro, on its second day. Moving young rays from the aquarium into a nursery tank is not ideal. However, this may become necessary if insufficient space is available and the parents start to attack the young. Photo: Hans Gonella

Export tanks at Turkys Aquarium in Manaus. Photo: bede-Verlag

dependent on their parents as they have to fend for themselves from day one. They will start feeding after a couple of days, although some young may have problems with that. It is obvious that they find it difficult to blow onto their food and then catch it with their mouths. If ample food is provided, they will have plenty to feed on and they learn quickly how to get to it. Although providing large amounts of food reduces the risk of losing some young, it does cause another problem. The young rays are extremely sensitive to even the smallest amounts of harmful substances in the water which uneaten food can produce. If that happens, the young will usually die in quick succession. Plenty of water changes can prevent this from happening. Good water filtration and of course the removal of any food left-overs from the aquarium after feeding is also important.

Live food should be the primary feed. Red mosquito larvae and *Tubifex* are normally happily accepted by the young. During the course of the first few weeks, other food types can be introduced. This can be the same food that the adult rays feed on. It must, however, be provided in smaller pieces so that the young – which have a smaller body diameter of 12 to 17cm – can swallow it without any problems.

For reasons of cleanliness the sandy substrate in the nursery tank should not be thicker than 1-2cm. This will enable the young rays to reach the food buried in the substrate and it will also be easy to remove any food remains with a vacuum siphon without extensively cleaning the substrate. During the first few weeks the nursery tank does not need to have any substrate at all. This will make it much easier to vacuum up any food remains from the floor. Later, when the young are feeding properly, the nursery aquarium will have to be provided with a substrate. After all, even young rays must also have the opportunity to bury themselves. If fed properly, young rays will double in size during the first year and they will grow quickly during subsequent years. For example, Peacock-eye Stingrays can attain a magnificent body diameter of 80cm before they are ten years old.

Keeping rays can be associated with many problems. The same health problems normally cause the death of most rays kept in domestic aquariums. It is ultimately down to the aquarist's experience, if rays fall ill. Failures are often blamed on the tap water or the bad health of newly acquired fish. The latter is only rarely the case. The most common cause for rays becoming ill is serious mismanagement of their care. Possible causes are tanks that are too small, poor water quality and inadequately fitted out aquariums. It is also possible that some unsuitable fittings in the tank can leak harmful substances. A diet lacking in variety can also have a detrimental effect on the rays' health.

The best way to protect against unpleasant surprises is to acquire healthy and strong rays. Unfavourable transport conditions or exposure to below-normal temperatures can lead to irreversible damage. These rays often refuse their food and over time slowly die. It is therefore important to check their state of health thoroughly before purchasing. Ideally you should watch the fish over several weeks in their tank in the aquatic shop to ensure that they feed properly before even considering buying them.

Unfortunately, "baby rays" are occasionally sold in aquatic shops. Such evidently small creatures, which have a body diameter of 8-12cm – sometimes a little more – often tempt buyers to an ill-considered purchase. The small size can give you the false impression that these

South American freshwater stingrays will live happily in a community tank with larger species of cichlids.
Photo: Hans Gonella

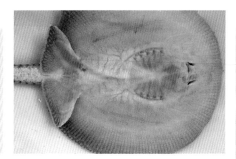

The underside of a freshwater stingray.
Photo: Dr Herbert R. Axelrod

rays can be kept in a small tank. Your joy over the new acquisition often doesn't last very long. The young rays are usually doomed to die. They quite often refuse their food and are otherwise weakened. It is not clear whether the reason for this is stress during transport or something else. One thing, however, is certain; caring for the young that do survive will prove to be difficult if they are kept in a small aquarium. So it is recommended that you only buy rays with a body diameter of 20-40cm. It goes without saying that an appropriately sized aquarium has to be made available for them.

> **Tip: As mentioned before, refusal of food is one of the problems that may arise when keeping rays.**

The reasons for food refusal are not always obvious. A monotonous diet and subsequent deficiency symptoms, for example, can lead to consistent food refusal. The same can happen after a prolonged enforced period of hunger. By the time that this problem has been recognized, the damage has already been done and any help often comes too late for the ray. Emaciated rays can rarely be rescued. A one-dimensional, protein-rich diet provided for the stingrays over a long period can be responsible for health problems. It is, therefore, important to provide a varied diet as well as an adequate supply of vegetable food as soon as you

The graceful, endlessly fascinating swimming movements of rays have been beautifully captured in this picture.
Photo: Yvette Tavernier

notice slimy faeces in the aquarium. An incorrect diet can lead to disturbance of growth. A regular and balanced diet is therefore essential. This applies particularly to young rays whose rapid growth depends on it.

Poor water quality as a cause of ill health

Rays react badly to poor water quality. This is usually the consequence of inappropriate care of freshwater stingrays. Often accompanied by abnormally fast breathing, the upper layer of skin, the epidermis, begins to disintegrate relatively quickly. The skin literally hangs from the body. At the same time affected rays display an increasingly listless behaviour and they die soon afterwards. Dead rays give off an ammoniac-like odour. Hardly anything else can be established. A number of reasons could be responsible for the demise of the rays, but nothing precise is known. Apart from water values, which are outside the required levels, high levels of nitrate in the aquarium are most likely to blame for the skin damage. Nitrate levels must never be higher than 50–70mg/l of water. As soon as you detect any skin deterioration on the rays, a partial water change must be carried out immediately and the nitrate levels in the water should be carefully checked. At the same time it is important to establish the reason for the deterioration in the quality of the water.

One possible reason could be a malfunctioning or dirty filtration system which is often accompanied by a massive drop of oxygen in the aquarium. Rays are also extremely sensitive to environmental poisons. For example, high copper levels leaching out from pipework can be harmful to rays. The only remedy is to let the water run through the pipes for a long time before carrying out a water change in order to keep copper levels as low as possible. However, this is usually only a problem with old installations. Unfortunately, even if you intervene immediately, any help that you offer usually comes too late and the illness cannot be cured. So it is important as a preventative measure to check the aquarium water on a regular basis and not to neglect your partial water changes and cleaning of the filtration system.

Infectious diseases

Diseases that most commonly affect aquarium fish are quite rare in freshwater stingrays. However, there are some infectious diseases which are known. But the aquarist will not always be able to find a cause for them. Red marks on the skin, for example, could be a **bacterial infection**, which may have been caused by an injury. If the infection does not clear up relatively quickly, the tissue will look bruised. In some cases the pectoral fin seams even curl up. Treating the infection with antibiotics may slow down the development of the illness but a cure is not necessarily guaranteed.

> **Tip: Rays may also suffer from parasites in the intestines. They become increasingly lethargic, stop feeding and eventually die.**

One indication of **intestinal parasites** can be slimy faeces. Threadworms can be the cause of this. If detected early, treatment with appropriate medication can be successful. **Fungal infection** is another ailment from which rays can suffer, usually as a consequence of an injury. These can be identified by a white layer or white thread-like patterns on the skin and they can also be treated with medication. Clean water conditions and the removal of anything in the aquarium which might cause an injury, for example sharp edges on rocks, can help to prevent fungal infections.

Admittedly, it is quite difficult to treat rays for infectious diseases. Quite often a treatment doesn't work as the causes of

the illnesses cannot be established. This can only be done by a pathological institute specializing in fish. Nonetheless, the treatment of diseased ray should not be considered a hopeless task, but remember that any medication should be used with caution. In particular, traditional remedies such as malachite green or methylene blue contain "copper" and are therefore less suitable as treatments. Neither should antibiotics be administered at random. They will end up in the water circulation where they can cause problems. It may be wiser to consult a vet who specializes in fish and then start treatment. However, this should be done at once as most diseases develop quickly in rays, and this leaves you little time to react.

Injuries

Quite often a ray's sting is covered with a piece of plastic pipe during transport. This is done to prevent rays injuring themselves or the transport containers getting damaged. This hose should remain on the sting when placing the ray into the aquarium. The earlier removal of the hose would be too stressful for the ray, as it may require some force to remove the protective cover. Furthermore, there is a substantial risk of the aquarist getting injured in the process. Therefore, you may decide simply to wait until the sting renews itself. However, failure to remove the sheath can lead to infection, so you must weigh up the risk of leaving the sheath in place against the hazards of manually removing it. But do be aware that removal is tricky and can be dangerous.

Rays often get injured during mating. However, these injuries usually heal very quickly so that no treatment is necessary. Provided the water is clean, there is no danger of any infections occurring. Neither should small injuries, inflicted by any of the fittings in the aquarium, have any serious consequences. But should these injuries occur repeatedly, then changes should be made.

Keeping different ray species in a community tank can often lead to problems. Occasionally two individuals from the same species may have a long-standing confrontation with one another. If these arguments continue over a long period, the weaker individual will get injured. Existing wounds will not heal because of repeated newly inflicted injuries and they can subsequently become infected. It may even happen that the weaker animal is literally torn to shreds and will die of its injuries. In such situations the antagonists must be separated.

It is almost impossible to treat such injuries. However, it is usually not necessary. But hygiene in the aquarium, such as partial water changes and the cleaning of the filtration system, becomes even more important when a ray is injured. Provided no germs are in the aquarium, the ray's own defences are quite sufficient to encourage the healing of a wound.

A ray in a transport basin.

Concluding remarks

It is doubtful whether the description of P. orbignyi *for this species of ray will gain acceptance. It is better known under the name* P. reticulatus. *Photo: Hans Gonella*

Keeping South American freshwater stingrays is certainly not a hobby that can be embarked on lightly in the spirit of an ill-considered adventure. To care for these fascinating fish is much more of a demanding job. People who cannot provide sufficiently large aquariums for rays, or who do not have sufficient time to care for them, must look for an alternative "easy-to-care-for" fish, however hard this decision may seem. After all, these magnificent fish can be enjoyed in the display aquariums of zoos where they are looked after expertly. In this way the responsible aquarist will make his indirect contribution towards the protection of natural habitats of freshwater stingrays.

The inexperienced aquarist should also abstain from keeping stingrays. Even though the vast amounts of specialist aquatic literature give a sound theoretical basis for fishkeeping, the specialized subject of keeping rays is much more demanding. This also means that not even the best reference book in the world can compensate for personal experience.

As in most cases only experienced aquarists concentrate on the care and breeding of South American freshwater stingrays, this book has not covered general basic aquatic knowledge. If you do not already possess this, it is absolutely essential to acquire this knowledge with the help of appropriate reference books. Keeping rays is one of those aquatic subjects which leaves a lot of questions unanswered. Many aspects remain to be studied in the future. This is what probably makes keeping rays so interesting, as it offers us an exciting opportunity to discover new things. Private groups of experts can therefore make a great scientific contribution through their personal commitment that helps fill the knowledge gaps about freshwater stingrays. And, in the final analysis, the relationship that one builds with the rays shows clearly how important it is to protect their natural habitats in the wild.

freshwater stingrays. This is especially important as over-development in South America is negatively influencing the

Books and articles in English

COMPAGNO, L.J.V. & COOK, S.F. 1995. Freshwater elasmobranchs: a questionable future. *Shark News*, 4.

ROSS, R. 1999. *Aqualog Special: Freshwater Stingrays from South America*. (Verlag A.C.S. GmbH).

ROSS, R. 1999. *Freshwater Stingrays*. (Barron's).

ROSS, R. 1998. *Freshwater Stingray Identification Guide*. Institute for Herpetological Research at the Santa Barbara Zoo, CA 93103, USA.

ROSS, R. & SCHAEFER, F. 2000. *Aqualog Special: Freshwater Rays*. (Verlag A.C.S. GmbH).

Books and articles in German

BAENSCH, H.A. & RIEHL, R. 1983, 1985, 1995. *Aquarien Atlas*, Band 1, 2 & 4. Melle.

CHAUMETTE, F. 1997. Aqua bizarr. Aquarium*live* 1 (4), 80.

DREYER, S. 1995. *Zierfische richtig füttern*. Ruhmannsfelden.

ELLIS, J. 1999. *Süsswasserstechrochen-Hybriden*. DATZ 52(2), 64.

FIEDLER, K. 1991. *Lehrbuch der Speziellen Zoologie*, Band II, Fische. Stuttgart.

GONELLA, H. 1996. Diskusfische im Rochenaquarium. *Diskus Jahrbuch Spezial* 3, 54-57.

GONELLA, H.1997. *Ratgeber Süsswasserrochen*. Ruhmannsfelden.

GONELLA, H. 1997. Nachzucht vom "Pfauenaugenstechrochen". Aquarium*live* 1(5), 11.

GONELLA, H. & SCHMIDT, J. 1999. Ecke der Importe. Süsswasserstechrochen – *Potamotrygon* sp. Aquarium*live* 3(2), 30-31.

GRZIMEK, B et al. 1970. Grzimeks Tierleben, *Enzyklopädie des Tierreiches*. Fische 1 (Vierter Band). Munich.

KADEN, J. 1980. "Fliegende Teppiche": Süsswasserrochen. *aquarien* magazin 14(9), 456-457.

KRAUSE, H.-J. 1990. *Handbuch Aquarienwasser*. Ruhmannsfelden.

KRAUSE, H.-J. 1994. *Handbuch Aquarientechnik*. Ruhmannsfelden.

LANGE, M. & SCHMIDT, J. 1998. Stachelrochenzucht. Aquarium*live* 2(3), 20.

LÜLING, K.-H. 1979. *Südamerikanische Fische und ihr Lebensraum*. Wuppertal-Elberfeld, Ettlingen.

SCHOBER, M. 1995. Südamerikanische Süsswasserstachelrochen im Aquarium. DATZ 48(5),285-287.

STAWIKOWSKI, R. 1993. Schwarze Rochen mit weissen Punkten. DATZ 46(1), 7-8.

STAWIKOWSKI, R. 1994. Die Fische Amazoniens. DATZ-Sonderheft, 19-20.

WICKER, R. 1991. Haltung und Zucht des Pfauenaugenstechrochens (*Potamotrygon motoro*). Jahresbericht des Zoologischen Gartens der Stadt Frankfurt a.M. 69-71.

WICKER, R 1998. Stachelrochen-Kinderschwemme. DATZ 51(8), 482.

Enjoy all the great titles in the AquaGuide series:

Aquarium Plants
ISBN 1-84286-034-8

Catfish
ISBN 1-84286-084-4

Discus
ISBN 1-84286-037-2

Freshwater Stingrays
ISBN 1-84286-083-6

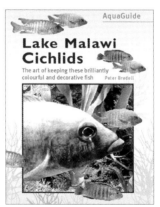

Lake Malawi Cichlids
ISBN 1-84286-035-6

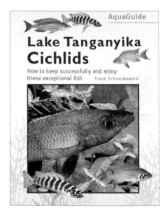

Lake Tanganyika Cichlids
ISBN 1-84286-036-4

For further information about these books and the full Interpet range of aquatic and pet titles, please write to:

Interpet Publishing, Vincent Lane, Dorking, Surrey, RH4 3YX
email: publishing@interpet.co.uk Tel: +44 (0)1306 873822